# Binary Quadratic Forms

Duncan A. Buell

# Binary Quadratic Forms

## Classical Theory and Modern Computations

Springer-Verlag
New York Berlin Heidelberg
London Paris Tokyo Hong Kong

Duncan A. Buell
Supercomputing Research Center
Bowie, MD 20715-4300, USA

Mathematical Subject Classification Codes: 11-02, 11R11, 11R29

Library of Congress Cataloging-in-Publication Data

Buell, Duncan A.
    Binary quadratic forms : classical theory and modern computations
/ Duncan A. Buell.
        p.   cm.
    Bibliography: p.
    1. Forms, Binary.   2. Forms, Quadratic.   1. Title.
QA201.B84   1989
512′.5—dc20                                              89-11314

Printed on acid-free paper.

Camera-ready copy prepared by the author using $T_EX$.
Printed and bound by Edwards Brothers, Incorporated, Ann Arbor, Michigan.
Printed in the United States of America.

9 8 7 6 5 4 3 2 1

ISBN 0-387-97037-1 Springer-Verlag New York Berlin Heidelberg
ISBN 3-540-97037-1 Springer-Verlag Berlin Heidelberg New York

# Preface

The first coherent exposition of the theory of binary quadratic forms was given by Gauss in the *Disquisitiones Arithmeticae*. During the nineteenth century, as the theory of ideals and the rudiments of algebraic number theory were developed, it became clear that this theory of binary quadratic forms, so elementary and computationally explicit, was indeed just a special case of a much more elegant and abstract theory which, unfortunately, is not computationally explicit. In recent years the original theory has been laid aside. Gauss's proofs, which involved brute force computations that can be done in what is essentially a two-dimensional vector space, have been dropped in favor of $n$-dimensional arguments which prove the general theorems of algebraic number theory. In consequence, this elegant, yet pleasantly simple, theory has been neglected even as some of its results have become extremely useful in certain computations.

I find this neglect unfortunate, because binary quadratic forms have two distinct attractions. First, the subject involves explicit computation and many of the computer programs can be quite simple. The use of computers in *experimenting* with examples is both meaningful and enjoyable; one can actually discover interesting results by computing examples, noticing patterns in the "data," and then proving that the patterns result from the conclusion of some provable theorem. The second attraction is that, although the theory of forms is just the quadratic case of results in algebraic number theory, the theorems can be independently proved (as in this book) using elementary methods. As Gauss recognized, it is truly delightful to build an elegant algebraic system from just the clever use of high-school algebra.

There have been a few other treatments of binary quadratic forms.

First, of course, there is Gauss. Then there is the book of G. B. Mathews, *Theory of Numbers,* published in 1896 but available in reprint form. Mathews is still the standard work on binary quadratic forms, although it predates the algebraic formulations of the early twentieth century and the invention and use of computers. Also, since the forms which are norm forms of quadratic number fields are written $(a, b, c)$, of discriminant $b^2 - 4ac$, whereas Gauss and Mathews discuss forms $(a, 2b, c)$ of determinant $b^2 - ac$, the notation of Gauss and Mathews must be considered "nonstandard"; theirs are not the forms with which the correspondence between the theory of forms and the theory of ideals can be directly made. There are also treatments in the Dirichlet-Dedekind *Zahlentheorie,* in Weber's *Lehrbuch der Algebra,* Volume III, in the texts of Landau and of Hecke, and occasional chapters in introductory texts in number theory.

My experience as a graduate student in mathematics, however, was that of frustration at every turn in trying to collect the computational theory of binary quadratic forms. Gauss, by modern standards, is difficult to read. Mathews is readable, but suffers from the notation problems–my own copy of the Chelsea reprint is liberally annotated with the insertion or removal of those annoying little 2's. It is my hope with this monograph to provide an update to Mathews' book (one book each century doesn't seem unnecessarily frequent), with modern notations and the "correct" definition of forms, an update which is a complete discussion of the classical theory of binary quadratic forms and also a survey of modern computations and applications. Since computations in any quantity are no longer done by hand but by machine, and since one of the major features of this theory is that clear and explicit examples of all theoretical constructs can be found, the emphasis, where possible, will be on explicit, algorithmically sensible proofs and examples. Constructive proofs will be preferred over existence proofs, and poor algorithms will be replaced by better algorithms where possible.

In spite of the notational complications, I must freely acknowledge my debt to the structure and in some places to the proofs of Mathews. I know of no other elementary treatment of this topic, so that of Mathews has become in my mind not just the only known way but the only *possible* way to proceed. I am also greatly indebted to Oliver Atkin, both for his inspiration and for the class notes I took from which this book in part derives.

This monograph requires some number theory as a prerequisite, perhaps half of a normal senior-level semester-long course, including the usual material on congruences, quadratic residues, and the reciprocity laws. In addition, "some" abstract algebra is necessary, again about half of a normal senior-level semester-long course. I use some basic group theory, homomorphisms, and describe some material for which the basics of ring theory, ideals, and fields would be useful but are not necessary. I state but do not prove the decomposition theorem for finite abelian groups so as to discuss Sylow subgroups and the special nature of class groups. Although a number of algebraic ideas are used, many of the more advanced or more difficult ideas are not necessary since the structures in number theory are, for the most part, abelian.

This is not intended as a classroom textbook since the material is not the canonical material of an introductory course in number theory or in algebra. It is intended, however, to be understandable by advanced undergraduates in mathematics or in computer science (and thus usable, perhaps, in a special "readings" course). Judging from the letters and questions I have received over the years from number theorists who understood the theoretical results but had not been exposed to the computational algorithms, I expect that my colleagues in number theory will also find it useful for its computational direction. My goal is a monograph that is readable without being pedantic, that covers completely the basic theory of binary quadratic forms, emphasizing computational and algorithmic aspects, and that adequately summarizes ongoing computations with forms, including both applications and the as-yet-unproved heuristics gained from the extensive computations of the last few years.

One of the more difficult parts of writing this book was to find the appropriate level of audience. I have tried not to demand too much background in algebra, and yet to mention the background where extremely relevant, so that naive readers will be given an appreciation of the connections with more advanced algebra and sophisticated readers who will recognize the existence of the connections will be satisfied that the precise theorems have been stated. I hope I have fulfilled that goal.

It is incumbent on me to thank those on whom early versions of this book were inflicted–my students at Louisiana State University, some of my colleagues at the Supercomputing Research Center, Marvin Wunderlich, Gary Cornell, Mary Ann Grandjean, and Walt Rudd. I am also indebted to Jerome Solinas and Robert L. Ward for detecting and correcting an error in the proof of Proposition 4.5 and to them and Michael J. Kascic, Jr., for reading this book with probably more care than I took in writing it.

What I *don't* know about mathematics is my fault. This book is dedicated to the three people who are largely responsible for that much smaller subset of what I *do* know about mathematics. They are David Buell, who first taught me the beauty of mathematics, John Brillhart, who first taught me modern algebra, and Oliver Atkin, who first taught me about binary quadratic forms.

# Contents

# Chapter 1

# Elementary Concepts

We consider here the *binary quadratic forms* in two variables

$$f(x,y) = ax^2 + bxy + cy^2$$

of *discriminant* $b^2 - 4ac = \Delta$. [1] The form can be written more simply
as $f = (a,\ b,\ c)$. It is often sufficient to write $(a,\ b,\ *)$ or $(a,\ *,\ *)$ if
the second or second and third coefficients are irrelevant or unknown or
are easily computed by the discriminant formula. Note that $\Delta \equiv 1$ or
$\Delta \equiv 0 \pmod 4$ is necessary and that $b \equiv \Delta \pmod 2$. Generally assume
also that we are dealing with *primitive* forms, for which $\gcd(a,b,c) = 1$.

A form $f(x,y)$ *represents* an integer $m$, positive, negative, or zero,
if there exist integers $x_0$, $y_0$ such that

$$ax_0^2 + bx_0y_0 + cy_0^2 = m.$$

The representation is *primitive* if $\gcd(x_0, y_0) = 1$. Among the early
questions asked about forms were the following:

---

[1] In this work it will almost always be the case that constants and variables are
rational integers. Exceptions to this general restriction should be obvious.

a) What integers can be represented by a given form?

b) What forms can represent a given integer?

c) If a form represents an integer, how many representations exist
and how may they all be found?

Many people in the seventeenth and eighteenth centuries, including
Fermat and Euler, sought to answer these questions and to classify
forms. The theory of forms was not put on solid footing, however,
until the publication of Gauss's *Disquisitiones Arithmeticae* in 1801.
Later, in the mid-nineteenth century, when the theory of ideals was
developed, it became clear that the theory of binary quadratic forms
was essentially identical with the theory of the class groups of quadratic
fields. The forms thus provide explicit and computationally convenient
examples, in the case of quadratic fields, of the elegant descriptions of
class groups of algebraic number fields in general.

As an example of representation we recall the following theorem
from elementary number theory, which concerns representation by the
form (1,0,1).

**Theorem 1.1.**

a) *If integers a and b exist such that the form $a^2 + b^2 = m$ is divisible
by any prime p of the form $4k - 1$, then $p \mid a$, $p \mid b$, and $p^2 \mid m$.*

b) *The only solutions to $x^2 + y^2 = 2$ are $(x, y) = (\pm 1, \pm 1)$, in which
the choices of sign are independent.*

c) *If p is any prime of the form $4k + 1$, then there exists a solution
$(x, y) = (a, b)$ to the equation $x^2 + y^2 = p$, and the only solutions*

*of this equation are* $(x, y) = (\pm a, \pm b)$ *and* $(\pm b, \pm a)$, *where the choices of sign are independent.*

d) *Let* $m = 2^t p_1^{r_1}...p_s^{r_s}$ *be a product of powers of primes, with the* $p_i$ *all being of the form* $4k + 1$. *Let* $\{(a_j, b_j)\}$ *be the solutions to* $x^2 + y^2 = p_j$ *for each* $j$. *Then, with* $i = \sqrt{-1}$, *all possible solutions to the equation* $x^2 + y^2 = m$ *are obtained by writing*

$$m = (1+i)^t (1-i)^t (a_1 + ib_1)^{r_1} (a_1 - ib_1)^{r_1}$$
$$\cdots$$
$$(a_s + ib_s)^{r_s} (a_s - ib_s)^{r_s}$$

*and then rewriting the right-hand side in any form* $(a + ib)(a - ib)$.

We also note that if we have

$$m = ax^2 + bxy + cy^2,$$

then we may complete the square to obtain

$$4am = (2ax + by)^2 - \Delta y^2.$$

Representation by forms thus subsumes representation as sums or differences of multiples of squares. We see immediately that for negative values of $\Delta$, the problem is finite and the number of representations must be finite. For positive values of $\Delta$, however, one representation will be shown to imply an infinite number of representations, all of which can easily be found from a "fundamental" representation.

Given some form $f$, consider the effect of the transformation

$$\begin{aligned} x &= \alpha x' + \beta y' \\ y &= \gamma x' + \delta y' \end{aligned} \tag{1.1}$$

where $\alpha$, $\beta$, $\gamma$, and $\delta$ are integers with $\alpha\delta - \beta\gamma \neq 0$. The form $f(x,y) = ax^2 + bxy + cy^2$ is transformed into

$$f'(x',y') = a'x'^2 + b'x'y' + c'y'^2,$$

where

$$
\begin{array}{rcl}
a' & = & a\alpha^2 + b\alpha\gamma + c\gamma^2 \\
b' & = & b(\alpha\delta + \beta\gamma) + 2(a\alpha\beta + c\gamma\delta) \\
c' & = & a\beta^2 + b\beta\delta + c\delta^2.
\end{array}
\tag{1.2}
$$

and

$$\Delta' = b'^2 - 4a'c' = (\alpha\delta - \beta\gamma)^2 \Delta.$$

We observe that $\Delta = \Delta'$ if and only if $\alpha\delta - \beta\gamma$ is $\pm 1$, and that inverting (1.1) results in

$$
x' = \frac{\delta x - \beta y}{\alpha\delta - \beta\gamma}
\tag{1.3}
$$

$$
y' = \frac{\alpha y - \gamma x}{\alpha\delta - \beta\gamma}
\tag{1.4}
$$

In (1.3) and (1.4), $f'$ is a form in integer variables if and only if the determinant $\alpha\delta - \beta\gamma$ of the transformation matrix is $\pm 1$. More specifically, we have the following:

a) the representation of an integer $m$ by $(a, b, c)$ with integer values $x$ and $y$ is equivalent to the representation of $m$ by $(a', b', c')$ with *integer* values $x'$ and $y'$ if and only if the determinant is $\pm 1$;

b) the representation of an integer $m$ by $(a, b, c)$ with integer values $x$ and $y$ implies the representation of $m$ by $(a', b', c')$ with *rational* values $x'$ and $y'$ regardless of the (nonzero) value of the determinant;

c) the representation of an integer $m$ by $(a, b, c)$ with integer values $x$ and $y$ is implied by the representation of $m$ by $(a', b', c')$ with integer values $x'$ and $y'$ regardless of the value of the determinant.

A form $f'(x', y') = (a', b', c')$ is defined to be *equivalent* to a form $f(x, y) = (a, b, c)$ if and only if $f'$ can be obtained from $f$ by a transformation of the form (1.1) for which $\alpha\delta - \beta\gamma = +1$, and we write $f \sim f'$. (Note the sign restriction.) We define here what is more precisely known as *proper equivalence*; forms obtained by transformations of determinant $-1$ are *improperly equivalent*. Since we shall only rarely use the notion of improper equivalence, the word "proper" shall not be used unless absolutely necessary. This is a true equivalence relation. For reflexivity, let $\alpha = \delta = 1$ and $\beta = \gamma = 0$. For symmetry, if

$$\begin{pmatrix} x \\ y \end{pmatrix} = \begin{pmatrix} \alpha & \beta \\ \gamma & \delta \end{pmatrix} \begin{pmatrix} x' \\ y' \end{pmatrix}$$

takes $f$ to $f'$, then

$$\begin{pmatrix} x' \\ y' \end{pmatrix} = \begin{pmatrix} \delta & -\beta \\ -\gamma & \alpha \end{pmatrix} \begin{pmatrix} x \\ y \end{pmatrix}$$

takes $f'$ to $f$. And if

$$\begin{pmatrix} x' \\ y' \end{pmatrix} = \begin{pmatrix} \varepsilon & \eta \\ \zeta & \theta \end{pmatrix} \begin{pmatrix} x'' \\ y'' \end{pmatrix}$$

then takes $f'$ to $f''(x'', y'')$, then

$$\begin{pmatrix} x \\ y \end{pmatrix} = \begin{pmatrix} \alpha & \beta \\ \gamma & \delta \end{pmatrix} \cdot \begin{pmatrix} \varepsilon & \eta \\ \zeta & \theta \end{pmatrix} = \begin{pmatrix} \alpha\varepsilon + \beta\zeta & \alpha\eta + \beta\theta \\ \gamma\varepsilon + \delta\zeta & \gamma\eta + \delta\theta \end{pmatrix} \begin{pmatrix} x'' \\ y'' \end{pmatrix}$$

takes $f$ to $f''$, giving transitivity.

It is important to point out the following inconvenience: if the transformation matrix $M$ takes $f$ to $f'$ and the matrix $M'$ takes $f'$ to $f''$, then the change of variables is done by the inverse:

$$\begin{pmatrix} x'' \\ y'' \end{pmatrix} = (M \cdot M')^{-1} \begin{pmatrix} x \\ y \end{pmatrix}.$$

For convenience in correlating with the more general definitions of quadratic forms, we present the matrix formulation of binary quadratic forms:

$$f(x,y) = (x\ y) \begin{pmatrix} a & b/2 \\ b/2 & c \end{pmatrix} \begin{pmatrix} x \\ y \end{pmatrix}.$$

Form equivalence is then matrix equivalence of the coefficient matrices, with

$$\begin{pmatrix} a' & b'/2 \\ b'/2 & c' \end{pmatrix} = \begin{pmatrix} \alpha & \gamma \\ \beta & \delta \end{pmatrix} \begin{pmatrix} a & b/2 \\ b/2 & c \end{pmatrix} \begin{pmatrix} \alpha & \beta \\ \gamma & \delta \end{pmatrix}.$$

The matrices involved in form equivalence are precisely the elements of the *classical modular group* $\Gamma$ of $2 \times 2$ matrices with integer coefficients and determinant $+1$. $\Gamma$ can be constructed as the free product on two generators

$$S = \begin{pmatrix} 1 & 1 \\ 0 & 1 \end{pmatrix},$$

for which

$$S^n = \begin{pmatrix} 1 & n \\ 0 & 1 \end{pmatrix},$$

and

$$T = \begin{pmatrix} 0 & -1 \\ 1 & 0 \end{pmatrix},$$

for which $T^2 = -I$. We will say more about this later.

Having defined an equivalence relation, we must next find canonical representatives for the equivalence classes. It will be determined that the number of classes of forms of a given discriminant $\Delta$, called the *class number* $h = h(\Delta)$, is always finite and that there are "natural" representatives for each class. Since it is the class and not (usually) the individual form which concerns us, we shall often say "form" when "class of forms" is meant. First, however, we must distinguish discriminants by sign.

If $\Delta$ is a perfect square, and a form $(a, b, c)$ of discriminant $\Delta$ represents $m$, then we can complete the square to obtain

$$4am = (2ax + by)^2 - \Delta y^2, \qquad (1.5)$$

whose right-hand side splits into

$$(2ax + by + y\sqrt{\Delta}) \cdot (2ax + by - y\sqrt{\Delta}).$$

This is the degenerate case, in which representation by a form and factoring into primes are almost indistinguishable. We assume, then, for the rest of this work, that $\Delta$ is not a perfect square. Conversely, if $m$ is zero, (1.5) shows that $\Delta$ must be a square, so we shall never consider representation of zero either.

If $\Delta < 0$, then since $b^2 - 4ac = \Delta$, $a$ and $c$ must have the same sign. For any integer $m$ represented by $f$, the completed square formula (1.5) shows that $m$ and $a$ (and $c$) have the same sign. We call forms of discriminant $\Delta < 0$ *definite* forms, and we choose to work only with the definite forms for which $a$ and $c$ are positive. These are the *positive definite* forms. If $\Delta > 0$, then $a$ and $c$ can be of opposite sign, and both positive and negative integers can be represented. These are, therefore, called *indefinite* forms.

A form $(k, kn, c)$ we shall call *ambiguous,* and the class of such a form an *ambiguous class.* [2] Such forms certainly exist for the odd $k$

---

[2] Gauss, writing in Latin, used the word *anceps*, meaning either "two-headed"

which divide $\Delta$, as is easily seen. For even $k$ not prime to $\Delta$, existence is only slightly more tedious to prove. We shall also refer to the forms $(a,\ b,\ a)$ as ambiguous, since the equivalence class to which they belong contains an ambiguous form: $b^2 - 4a^2 = \Delta$ means that $(b+2a)(b-2a) = \Delta$. So

$$(a,\ b,\ a) \sim (a,\ b+2a,\ b+2a) \sim$$
$$(b+2a,\ -b-2a,\ a) \sim (b+2a,\ b+2a,\ a).$$

These will turn out to be valuable forms in the future.

---

or "ambiguous," and with deliberate reference to the Roman god Janus. French translations have usually favored the word *ambigu*, but in German both *ambige* and *zweiseitig* ("two-sided") have been used.

# Addendum–The Modular Group

**Theorem 1.2.** *Let $\Gamma$ be the classical modular group, that is, the group of $2 \times 2$ matrices with integer coefficients and determinant $+1$. $\Gamma$ is the free product group on the generators*

$$S = \begin{pmatrix} 1 & 1 \\ 0 & 1 \end{pmatrix}$$

*and*

$$T = \begin{pmatrix} 0 & -1 \\ 1 & 0 \end{pmatrix}.$$

*That is, any matrix $M \in \Gamma$ can be written as*

$$S^{i_1} T^{j_1} S^{i_2} \cdots S^{i_k} T^{j_k}$$

*with the $i_l$ and $j_l$ integers. $\Gamma$ is also the free product on the generators $T$ and $P = T^3 S$. The relations $T^4 = P^3 = I$ are defining relations for $\Gamma$.*

**Proof.** It is clear that $T$ and $P$ are of orders 3 and 4, respectively, and that the inverse of

$$S^n = \begin{pmatrix} 1 & n \\ 0 & 1 \end{pmatrix}$$

is

$$S^{-n} = \begin{pmatrix} 1 & -n \\ 0 & 1 \end{pmatrix}.$$

The proof that $\Gamma$ is free on $T$ and $S$ closely resembles the Euclidean algorithm. A given element of $\Gamma$ is multiplied on the left by powers of $S$ and $T$ until the identity is obtained. Let $M \in \Gamma$ be any such matrix. Then one of $M, TM, T^2M,$ or $T^3M$ is a matrix

$$\begin{pmatrix} \alpha & \beta \\ \gamma & \delta \end{pmatrix}$$

for which $\beta > 0$ and $\beta \geq | \delta |$. If $\delta$ is not zero, we multiply on the left

by $S^n$ to get

$$\begin{pmatrix} \alpha + \gamma n & \beta + \delta n \\ \gamma & \delta \end{pmatrix}.$$

We choose $n$ so that we have $| \delta | > \beta + \delta n \geq 0$. Continuing this way, we

eventually obtain a matrix with one of the upper right or lower right

entries equal to zero. Applying $T$ if necessary, we obtain

$$\begin{pmatrix} \alpha & 0 \\ \gamma & \delta \end{pmatrix}.$$

Since determinants are multiplicative, we have $\alpha = \delta = \pm 1$. Multiply-

ing by $T^2$ if necessary, we obtain the matrix

$$\begin{pmatrix} 1 & 0 \\ \gamma & 1 \end{pmatrix}.$$

Multiplying now by $T^3 S^\gamma T$, we obtain the identity, as desired.

That $\Gamma$ is freely generated by $T$ and $P$ follows from the fact that

it is freely generated by $T$ and $S$. We must show now that no other

independent relations hold in $\Gamma$. If we had a relation

$$T^{a_1} P^{b_1} T^{a_2} P^{b_2} \cdots T^{a_n} P^{b_n} = I \qquad (1.6)$$

for integers $a_i$ and $b_i$, then since $T^2$ commutes with all elements of $\Gamma$,

we may reduce (1.6) to

$$P^{b_1} T P^{b_2} \cdots T P^{b_n} T = T^c$$

with each $b_i$ either 1 or 2 and $c = 0, 1, 2,$ or 3. However, both $PT$ and

$P^2 T$ are matrices $\begin{pmatrix} \alpha & \beta \\ \gamma & \delta \end{pmatrix}$ for which we have either $\alpha, -\beta, -\gamma, \delta \leq 0$
and $\beta + \gamma > 0$ or $\alpha, -\beta, -\gamma, \delta \geq 0$ and $\beta + \gamma < 0$; this property is

maintained upon multiplication on the right by any element of $\Gamma$, but since none of the matrices $T^c$ have this property, no relation (1.6) can exist.

The modular group $\Gamma$ can be considered as a group of linear fractional transformations of the complex upper half plane

$$\mathbf{H} = \{z = x + y\sqrt{-1} \in \mathbf{C} : y \geq 0\}$$

onto itself, with the matrix

$$M = \begin{pmatrix} \alpha & \beta \\ \gamma & \delta \end{pmatrix}$$

corresponding to the mapping

$$z \mapsto \frac{\alpha z + \beta}{\gamma z + \delta}.$$

In this correspondence, the group operation is clearly preserved.

For any group $G$ of one-to-one transformations of a set $X$ to itself, a *fundamental domain* $F$ (if one exists) is a subset of $X$ such that any point in $X$ can be mapped by some transformation in $G$ to some point in $F$, and no two points in the interior of $F$ can be mapped to each other by any transformation in $G$.

$\Gamma$ is generated as mappings as the free product of $S(z) = z + 1$ and $T(z) = -1/z$. A fundamental domain $F$ for $\Gamma$ is the set

$$\begin{aligned} F = \quad & \{z \in \mathbf{H} : \mid z \mid > 1, -1/2 < \Re(z) < 1/2\} \cup \\ & \{z \in \mathbf{H} : \mid z \mid \geq 1, \Re(z) = -1/2\} \cup \\ & \{z \in \mathbf{H} : \mid z \mid = 1, \Re(z) \leq 0\}. \end{aligned}$$

Mimicking the proof that $\Gamma$ is generated by $S$ and $T$ reveals that any point in $\mathbf{H}$ is equivalent to some point in $F$ under the action of $\Gamma$, and that no points in $F$ are equivalent except $i = T(i)$ and

$$\frac{-1 + \sqrt{-3}}{2} = T\left(S\left(\frac{-1 + \sqrt{-3}}{2}\right)\right).$$

The special nature of these two points will become more apparent in the next chapter.

# Chapter 2

# Reduction of Positive Definite Forms

First to be considered are positive definite forms, that is, forms with negative discriminants $\Delta = -D$. The primary goal of this chapter is to determine a canonical representative, called the reduced form, for each equivalence class of forms of negative discriminant. We define $f = (a,\ b,\ c)$ to be *reduced* if

$$|\ b\ | \le a \le c. \tag{2.1}$$

**Proposition 2.1.** *If $f = (a,\ b,\ c)$ is a reduced form of discriminant $-D$, then $|\ b\ | \le \sqrt{D/3}$.*

**Proof.** $4b^2 \le 4ac = b^2 + D$, so $3b^2 \le D$.

**Theorem 2.2.** *The number of reduced forms of a fixed discriminant $-D$ is finite.*

**Proof.** There are only finitely many candidates for reduced forms since by Proposition 2.1 the set of possible $b$'s is finite, and each such $b$ determines a finite set of factorings of $b^2 + D$ into $4ac$.

**Theorem 2.3.** *Every form $f$ of discriminant $-D$ is equivalent to a reduced form of the same discriminant.*

**Proof.** We give the standard reduction algorithm. Let $(a, b, c)$ be a form of discriminant $-D$. If this form is not reduced, an integer $\delta$ can be chosen such that $|-b + 2c\delta| \leq |c|$. Then

$$(a, b, c) \sim (c, -b + 2c\delta, a - b\delta + c\delta^2) = (a', b', c').$$

We now have $|b'| \leq a'$. If $a' \leq c'$, we are done. If not, we repeat the process. Since we only continue the reduction when $c' < a' = c$, and since these are positive integers, the process must terminate, yielding a reduced form.

*Remarks.*

a) The matrix for the reduction transformation is important:

$$\begin{pmatrix} 0 & -1 \\ 1 & \delta \end{pmatrix}$$

gives

$$(a, b, c) \sim (c, -b + 2c\delta, a - b\delta + c\delta^2). \qquad (2.2)$$

In practice, $c'$ is usually computed as $(b'^2 + D)/(4a')$.

b) The standard reduction algorithm can be considered as a single reduction transformation. It is not the most efficient method for reduction, however. If $a$ is less than $c$, but $b$ is too large for a reduced form, the initial reduction transformation will be to transform the form $(a, b, c)$ to the equivalent form $(c, -b, a)$ $(\delta =$

0) before applying a transformation that reduces the magnitude of $b$. For hand calculation, then, in this case, the reduction step $(a,\ b,\ c) \sim (a,\ b + 2a\beta,\ a\beta^2 + b\beta + c)$ should be applied with $\beta$ chosen so as to make $b + 2a\beta$ smaller in magnitude than $a$. This is the reduction matrix

$$\begin{pmatrix} 1 & \beta \\ 0 & 1 \end{pmatrix}.$$

c) From a computational point of view, the similarity between reduction of forms and a greatest common divisor operation should be obvious.

**Theorem 2.4.** *With the exception of*

$$1)\quad (a,\ b,\ a) \sim (a,\ -b,\ a)$$
$$2)\quad (a,\ a,\ c) \sim (a,\ -a,\ c)$$

*no distinct reduced forms are equivalent.*

**Proof.** If $(a,\ b,\ c)$ and $(a',\ b',\ c')$ are equivalent and reduced, then $a' = a\alpha^2 + b\alpha\gamma + c\gamma^2$ for some $\alpha, \gamma$. Then, since $a \geq a'$ (or we would reverse everything), we have

$$a \ \geq a\alpha^2 + b\alpha\gamma + c\gamma^2 \geq a(\alpha^2 + \gamma^2) + b\alpha\gamma$$
$$\geq a(\alpha^2 + \gamma^2) - a \mid \alpha\gamma \mid \geq a \mid \alpha\gamma \mid.$$

We conclude that the only allowable cases are $(\alpha, \gamma) = (0, \pm 1), (\pm 1, 0)$, and $(\pm 1, \pm 1)$, where the signs in the last are independent.

*Case 1.* $\alpha = \pm 1$, $\gamma = 0$. Then $(a, b, c) \sim (a, b + 2a\beta, *)$ or $(a, b - 2a\beta, *)$. The only way, however, to have $| b | \leq a$ and $| b \pm 2a\beta | \leq a$ is to have $\beta = 0$ (in which case the forms are identical) or to have $\beta = \pm 1$ as appropriate and $(a, a, c) \sim (a, -a, c)$.

*Case 2.* $\alpha = 0$, $\gamma = \pm 1$. Then $(a, b, c) \sim (c, -b \pm 2c\delta, *)$ are equivalent reduced forms. But the center coefficients cannot both be small enough unless $\delta = 0, -1, +1$. If $\delta = 0$, then to have $(a, b, c)$ and $(c, -b, a)$ both be reduced, we must have $c = a$ and thus $(a, b, a) \sim (a, -b, a)$. If $\delta = \pm 1$, then $(a, c, c) \sim (c, -c, a)$, but since $a \leq c$, and $c \leq a$, we really have $(a, a, a) \sim (a, -a, a)$.

*Case 3.* $\alpha = \pm 1$, $\gamma = \pm 1$ independently. In this case we have $a \geq a' \geq a | \alpha\gamma | = a$, so $a = a' = a \pm b + c$. This means $(a', b', c') = (a, b, \pm b)$, but since the form is reduced, we have $a = \pm b$. The reduced equivalent forms are again $(a, a, a)$ and $(a, -a, a)$.

*Remark.* The form $(a, 0, a)$ is equivalent to itself under the transformations $\pm T$ and that $(a, a, a)$ is equivalent to itself under $\pm P$ and $\pm P^2$, where

$$P = TS = \begin{pmatrix} 0 & -1 \\ 1 & 1 \end{pmatrix}.$$

These nontrivial self-equivalences of a form are called *automorphs* of the form. The above argument, carefully studied, shows that no other reduced forms possess automorphs. Automorphs will be closely studied with regard to indefinite forms in Chapter 3.2, since for those forms there are nontrivial automorphs.

*Examples.*

The equivalence $(a, b, a) \sim (a, -b, a)$ under $x' = -y$ and $y' = x$ is easy to see: $ax^2 + bxy + ay^2$ is symmetric in $x$ and $y$, and changing $b$ to

$-b$ and changing the sign of one of the variables leave the representation unchanged.

The equivalence $(a,\ a,\ c) \sim (a,\ -a,\ c)$ under $x' = x - y$ and $y' = y$ is similarly easy. Completing the square in any representation yields

$$4am = (2ax + ay)^2 + Dy^2.$$

Under the equivalence, the second summand on the right is unchanged since $y = y'$, and so we need only look at the first summand:

$$2ax + ay = 2a(x' - y') + ay' = 2a'x' - 2a'y' + a'y' = 2a'x' - a'y'.$$

In either of the two exceptional cases above, the equivalence class representative is chosen so as to have a nonnegative center coefficient. With this choice the following major theorems are clear.

**Theorem 2.5.** *Every form of discriminant $-D < 0$ is equivalent to a unique [given the above convention] reduced form.*

**Theorem 2.6.** *The number of equivalence classes for a given discriminant is finite.*

We next define the reduced form $(1,\ 1,\ (D-1)/4)$ or $(1,\ 0,\ D/4)$, depending on the parity of $D$, to be the *principal form*. (Remember that all discriminants $-D$ must be congruent to either 0 or to 1 modulo 4.)

The equivalence and reduction of positive definite forms has the following geometric interpretation. In Chapter 1 the *fundamental domain* $F$ for $\Gamma$ was given as the set

$$F = \{z \in \mathbf{H} : |z| > 1, -1/2 < \Re(z) < 1/2\}.$$

Given a form $f = (a, b, c)$ of discriminant $-D$, we define the *principal root* of $f$ to be the complex number

$$\tau = \frac{-b + \sqrt{-D}}{2a}.$$

Note that $\tau$ is one of the solutions to $ax^2 + bx + c = 0$.

**Proposition 2.7.** *A form is reduced if and only if its principal root lies in the closure of $F$.*

**Proof.** The condition $\mid b \mid \le a$ is equivalent to a condition that $-1/2 \le \Re(\tau) \le 1/2$. The condition $a \le c$ is equivalent to

$$1 \le \frac{c}{a} = \frac{4ac}{4a^2} = \frac{b^2 + D}{4a^2} = \frac{(-b + \sqrt{-D}) \cdot (-b - \sqrt{-D})}{2a \cdot 2a} = \tau\tau' = \mid \tau \mid,$$

where $\tau'$ is the complex conjugate of $\tau$.

Eliminating the $\Re(z) = 1/2$ boundary of $F$ is equivalent to choosing a form representative with positive center coefficient, if $\mid b \mid = a$, and the forms $(a, 0, a)$ and $(a, a, a)$, the only forms which are nontrivially automorphic to themselves, correspond respectively to principal roots $\tau = \sqrt{-1}$ and $\tau = (-1 + \sqrt{-3})/2$. These are the special points of the fundamental domain mentioned in the Appendix to Chapter 1. A geometric technique for reduction, then, would be

a) Apply $S^n$ or $S^{-n}$ as necessary to get $\mid \Re(\tau) \mid \le 1/2$.

b) Apply $T$ as necessary to get $\mid \tau \mid \ge 1$.

c) Repeat as necessary.

**Examples**

| $\Delta$ | $h$ | Reduced Form Representatives of Classes |
|---|---|---|
| $-3$ | 1 | $(1,1,1)$ |
| $-7$ | 1 | $(1,1,2)$ |
| $-11$ | 1 | $(1,1,3)$ |
| $-15$ | 2 | $(1,1,4),(2,1,2)$ |
| $-19$ | 1 | $(1,1,5)$ |
| $-23$ | 3 | $(1,1,6),(2,\pm1,3)$ |
| $-27$ | 1 | $(1,1,7)$ |
| $-31$ | 3 | $(1,1,8),(2,\pm1,4)$ |
| $-35$ | 2 | $(1,1,9),(3,1,3)$ |
| $-39$ | 4 | $(1,1,10),(2,\pm1,5),(3,3,4)$ |
| $-43$ | 1 | $(1,1,11)$ |
| $-47$ | 5 | $(1,1,12),(2,\pm1,6),(3,\pm1,4)$ |
| $-51$ | 2 | $(1,1,13),(3,3,5)$ |
| $-55$ | 4 | $(1,1,14),(2,\pm1,7),(4,3,4)$ |
| $-59$ | 3 | $(1,1,15),(3,\pm1,5)$ |
| $-63$ | 4 | $(1,1,16),(2,\pm1,8),(4,1,4)$ |
| $-67$ | 1 | $(1,1,17)$ |
| $-71$ | 7 | $(1,1,18),(2,\pm1,9),(3,\pm1,6),(4,\pm3,5)$ |
| $-75$ | 2 | $(1,1,19),(3,3,7)$ |
| $-79$ | 5 | $(1,1,20),(2,\pm1,10),(4,\pm1,5)$ |
| $-83$ | 3 | $(1,1,21),(3,\pm1,7)$ |
| $-87$ | 6 | $(1,1,22),(2,\pm1,11),(3,3,8),(4,\pm3,6)$ |
| $-91$ | 2 | $(1,1,23),(5,3,5)$ |
| $-95$ | 8 | $(1,1,24),(2,\pm1,12),(3,\pm1,8),(4,\pm1,6),(5,5,6)$ |
| $-99$ | 2 | $(1,1,25),(5,1,5)$ |
| $-103$ | 5 | $(1,1,26),(2,\pm1,13),(4,\pm3,7)$ |
| $-107$ | 3 | $(1,1,27),(3,\pm1,9)$ |
| $-111$ | 8 | $(1,1,28),(2,\pm1,14),(3,3,10),4,\pm1,7),(5,\pm3,6)$ |
| $-115$ | 2 | $(1,1,29),(5,5,7)$ |
| $-119$ | 10 | $(1,1,30),(2,\pm1,15),(3,\pm1,10),(4,\pm3,8),(5,\pm1,6),(6,5,6)$ |
| $-123$ | 2 | $(1,1,31),3,3,11)$ |
| $-127$ | 5 | $(1,1,32),(2,\pm1,16),(4,\pm1,8)$ |
| $-131$ | 5 | $(1,1,33),(3,\pm1,11),(5,\pm3,7)$ |
| $-135$ | 6 | $(1,1,34),(2,\pm1,17),(4,\pm1,9),(5,5,8)$ |
| $-139$ | 3 | $(1,1,35),(5,\pm1,7)$ |
| $-143$ | 10 | $(1,1,36),(2,\pm1,18),(3,\pm1,12),(4,\pm1,9),(6,1,6),(6,\pm5,7)$ |
| $-147$ | 2 | $(1,1,37),(3,3,13)$ |
| $-151$ | 7 | $(1,1,38),(2,\pm1,19),(4,\pm1,10),(5,\pm1,8)$ |
| $-155$ | 4 | $(1,1,39),(3,\pm1,13),(5,5,9)$ |
| $-159$ | 10 | $(1,1,40),(2,\pm1,20),(3,3,14),(4,\pm1,10),(5,\pm1,8),(6,\pm3,7)$ |
| $-163$ | 1 | $(1,1,41)$ |

| $\Delta$ | $h$ | Reduced Form Representatives of Classes |
|---|---|---|
| $-4$ | 1 | $(1,0,1)$ |
| $-8$ | 1 | $(1,0,2)$ |
| $-12$ | 1 | $(1,0,3)$ |
| $-16$ | 1 | $(1,0,4)$ |
| $-20$ | 2 | $(1,0,5),(2,2,3)$ |
| $-24$ | 2 | $(1,0,6),(2,0,3)$ |
| $-28$ | 1 | $(1,0,7)$ |
| $-32$ | 2 | $(1,0,8),(3,2,3)$ |
| $-36$ | 2 | $(1,0,9),(2,2,5)$ |
| $-40$ | 2 | $(1,0,10),(2,0,5)$ |
| $-44$ | 3 | $(1,0,11),(3,\pm2,4)$ |
| $-48$ | 2 | $(1,0,12),(3,0,4)$ |
| $-52$ | 2 | $(1,0,13),(2,2,7)$ |
| $-56$ | 4 | $(1,0,14),(2,0,7),(3,\pm2,5)$ |
| $-60$ | 2 | $(1,0,15),(3,0,5)$ |
| $-64$ | 2 | $(1,0,16),(4,4,5)$ |
| $-68$ | 4 | $(1,0,17),(2,2,9),(3,\pm2,6)$ |
| $-72$ | 2 | $(1,0,18),(2,0,9)$ |
| $-76$ | 3 | $(1,0,19),(4,\pm2,5)$ |
| $-80$ | 4 | $(1,0,20),(3,\pm2,7),(4,0,5)$ |
| $-84$ | 4 | $(1,0,21),(2,2,11),(3,0,7),(5,4,5)$ |
| $-88$ | 2 | $(1,0,22),(2,0,11)$ |
| $-92$ | 3 | $(1,0,23),(3,\pm2,8)$ |
| $-96$ | 4 | $(1,0,24),(3,0,8),(4,4,7),(5,2,5)$ |
| $-100$ | 2 | $(1,0,25),(2,2,13)$ |

# Chapter 3

# Indefinite Forms

## 3.1 Reduction, Cycles

We now consider the indefinite forms, that is, the forms of positive discriminant $\Delta = D > 0$. Our treatment will closely follow that of Mathews, our goal again being the determination of canonical forms for the equivalence classes. In the case of negative discriminants, the "reduced" forms are essentially unique in a given equivalence class. For positive discriminants, however, it is not only the case that many reduced forms can lie in the same class, an elegant structure is possessed by the reduced forms–they form cycles. An indefinite form $(a, \ b, \ c)$ of discriminant $D > 0$ is called *reduced* if

$$0 < b < \sqrt{D} \tag{3.1}$$
$$\sqrt{D} - b < 2 \mid a \mid < \sqrt{D} + b$$

We make several easy deductions.

**Proposition 3.1.** *If $(a, \ b, \ c)$ is reduced, then $\sqrt{D} - b < 2 \mid c \mid < \sqrt{D} + b$.*

**Proof.** Since $b^2 - 4ac = D$, we have

$$(\sqrt{D} - b) \cdot (\sqrt{D} + b) = -4ac = (2 \mid a \mid) \cdot (2 \mid c \mid).$$

Since $0 < b < \sqrt{D}$, $\sqrt{D} - b < \sqrt{D} + b$. We have the situation $xy = zw$, with $x < z < y$, and it follows that $x < w < y$.

**Proposition 3.2.** *The number of reduced forms of a given discriminant is finite.*

**Proof.** The number of values for $b$ has been limited, so the finite number of reduced forms follows from the finite number of factorings of $b^2 - D$ into $4ac$.

**Proposition 3.3.** *Any indefinite form is equivalent to a reduced form of the same discriminant.*

**Proof.** We give a reduction algorithm. If $(a, b, c)$ is not reduced, we choose $\delta$ (which in this case is necessarily unique) such that

$$\sqrt{D} - 2 \mid c \mid < -b + 2c\delta < \sqrt{D},$$

and we have

$$(a, b, c) \sim (c, -b + 2c\delta, a - b\delta + c\delta^2)$$

If $\mid a - b\delta + c\delta^2 \mid < \mid c \mid$, the process is repeated. As in the reduction of definite forms, the reduction process must be finite, terminating when we get a form $(A, B, C)$ such that $\mid A \mid \leq \mid C \mid$ and $\sqrt{D} - 2 \mid A \mid < B < \sqrt{D}$. If this is true, then $\sqrt{D} - B < 2 \mid A \mid$. Further, since

$$\mid \sqrt{D} - B \mid \cdot \mid \sqrt{D} + B \mid = 4 \mid A \mid \mid C \mid,$$

we must then have $| \sqrt{D} + B | > 2 | C |$. We continue the inequality:

$$| \sqrt{D} + B | > 2 | C | > 2 | A | > \sqrt{D} - B.$$

Looking at the left and right ends of this, we see that $B$ must be positive, so that $0 < B < \sqrt{D}$ and $(A, B, C)$ is reduced.

We define two reduced forms $(a, b, a')$ and $(a', b', c')$ to be *adjacent* if $b + b' \equiv 0 \pmod{2a'}$. It is easy to see that there is a unique reduced form adjacent to the right and to the left of any given reduced form.

Once again, there is a strong computational similarity between the reduction algorithm and the standard algorithm for the greatest common divisor. As will be seen later in this chapter, more than a mere similarity exists. Reduction of definite forms is identical with the continued fraction expansion of a related quadratic irrational, and the continued fraction algorithm applied to a rational number is precisely the Euclidean algorithm.

**Proposition 3.4.** *The set of reduced forms of a given discriminant can be partitioned into cycles of adjacent forms.*

**Proof.** We begin with any reduced form and proceed to the right through successively adjacent reduced forms. Since the set of reduced forms is finite, the list of successively adjacent forms must return to the original form. If there are no more reduced forms, the process is finished; otherwise, we choose a form not yet used and repeat the process.

Since adjacent forms are equivalent, under the matrix transformation

$$\begin{pmatrix} 0 & -1 \\ 1 & \frac{b+b'}{2a} \end{pmatrix},$$

and equivalence is transitive, all forms in a given cycle are equivalent
to each other.

Proposition 3.4 is the easy half of the following major theorem. The
difficult half of the proof will be presented in Section 3.3 so as not to
disturb the continuity of the discussion.

**Theorem 3.5.** *Two reduced forms are equivalent if and only if they
are in the same cycle.*

We call the form $(a, -b, c)$ the *opposite* of the form $(a, b, c)$. An
ambiguous form is equivalent to its own opposite, since if $b = ka$, the
choice $\delta = k$ gives

$$(a, b, c) \sim (c, -b, a) \sim (a, b - 2a\delta, c - b\delta + a\delta^2) = (a, b, c).$$

We further define forms $(a, b, c)$ and $(c, b, a)$ to be *associated*. We
note that opposite forms are improperly equivalent (obtainable one
from another by a matrix transformation of determinant $-1$) under

$$\begin{pmatrix} 1 & 0 \\ 0 & -1 \end{pmatrix}$$

and its negative, and associated forms are improperly equivalent under

$$\begin{pmatrix} 0 & 1 \\ 1 & 0 \end{pmatrix}$$

and its negative.

**Proposition 3.6.** *The number of forms in any cycle, called the period
of the cycle, is always even.*

**Proof.** The first and last coefficients of any reduced form are of opposite sign. We may therefore form pairs of adjacent forms $(a, b, c) \sim$

$(c,\ b',\ c')$ in which the coefficient $c$ is negative and $a$ and $c'$ are positive. Because the adjacency is clearly an adjacency of these pairs, it takes an integral number of pairs to form any cycle.

**Proposition 3.7.** *If the form $f'$, associate to $f$, is in a different cycle from that of $f$, then this is true for all forms in both cycles, which we call associated cycles.*

**Proof.** Cycling forward (to the right) from $f$, the form adjacent to $f = (a,\ b,\ c)$ is $(c,\ b',\ a')$. Cycling backward (to the left) from $f'$ yields $(a',\ b',\ c) \sim (c,\ b,\ a)$. That is, cycling forward from $f$ we encounter the associates of the forms encountered when cycling backward from $f'$.

**Proposition 3.8.** *A cycle which contains any ambiguous form contains exactly two and is its own associate. Conversely, a cycle which is its own associate contains exactly two ambiguous forms.*

**Proof.** If a form $f$ and its associate $f'$ are in the same cycle, then we can cycle forward from $f$ and backward from $f'$ through pairs of associated forms. Since the cycles have finite length, we must eventually arrive at adjacent associated forms $(a',\ b,\ a) \sim (a,\ b,\ a')$. Since these are adjacent, we have $b + b \equiv 0 \pmod{2a}$; that is, $a|b$, so that $(a,\ b,\ a')$ is ambiguous. Similarly, cycling backward from $f$ and forward from $f'$ will produce a different ambiguous form. A self-associate cycle thus contains two ambiguous forms. It cannot contain more since the cycle is complete when the second ambiguous form is found. And it is easy to see that a cycle which contains an ambiguous form must be self-associate since the form $(a,\ ak,\ c)$ is the form adjacent to its own associate $(c,\ ak,\ a)$.

We call the reduced form $(1, b, c)$ the *principal form* for a given discriminant, and the cycle in which it lies the *principal cycle*.

# General Examples and Observations

For negative discriminants, reduced forms are, in general, asymmetric since the third coefficient is at least as large as the first. For positive discriminants, this is not true. Indeed, reduced forms occur in groups: for any given lead coefficient $a$ the existence of one reduced form $(a, b, c)$ implies the existence of the reduced forms $(a, b, c)$, $(-a, b, -c)$, $(c, b, a)$, and $(-c, b, -a)$. Further, since solutions to $b^2 \equiv D \pmod{a}$ occur in pairs, we also have reduced forms $(a, -b + 2a\sigma, a - b\sigma + c)$, $(-a, -b + 2a\sigma, -a + b\sigma - c)$, $(a - b\sigma + c, -b + 2a\sigma, a)$, and $(-a + b\sigma - c, -b + 2a\sigma, -a)$, where $\sigma$ is the sign of $a$. These generally lead to further forms, and so on. The following examples of cycles will illustrate the previous discussion.

For $D = 1173 = 3 \cdot 17 \cdot 23$ there are four cycles:

A) $(1, 33, -21) \sim (-21, 9, 13) \sim (13, 17, -17) \sim (-17, 17, 13) \sim$
$(13, 9, -21) \sim (-21, 33, 1)$

B) $(-1, 33, 21) \sim (21, 9, -13) \sim (-13, 17, 17) \sim (17, 17, -13) \sim$
$(-13, 9, 21) \sim (21, 33, -1)$

C) $(3, 33, -7) \sim (-7, 23, 23) \sim (23, 23, -7) \sim (-7, 33, 3)$

D) $(-3, 33, 7) \sim (7, 23, -23) \sim (-23, 23, 7) \sim (7, 33, -3)$

For $D = 1313 = 13 \cdot 101$ there are also four cycles:

A) $(1, 35, -22) \sim (-22, 9, 14) \sim (14, 19, -17) \sim (-17, 15, 16) \sim$
$(16, 17, -16) \sim (-16, 15, 17) \sim (17, 19, -14) \sim (-14, 9, 22) \sim$

$(22, 35, -1) \sim (-1, 35, 22) \sim (22, 9, -14) \sim (-14, 19, 17) \sim$
$(17, 15, -16) \sim (-16, 17, 16) \sim (16, 15, -17) \sim (-17, 19, 14) \sim$

$(14, 9, -22) \sim (-22, 35, 1)$

B) $(13, 13, -22) \sim (-22, 31, 4) \sim (4, 33, -14) \sim (-14, 23, 14) \sim$
$(14, 33, -4) \sim (-4, 31, 22) \sim (22, 13, -13) \sim (-13, 13, 22) \sim$
$(22, 31, -4) \sim (-4, 33, 14) \sim (14, 23, -14) \sim (-14, 33, 4) \sim$
$(4, 31, -22) \sim (-22, 13, 13)$

C) $(7, 23, -28) \sim (-28, 33, 2) \sim (2, 35, -11) \sim (-11, 31, 8) \sim$
$(8, 33, -7) \sim (-7, 23, 28) \sim (28, 33, -2) \sim (-2, 35, 11) \sim$
$(11, 31, -8) \sim (-8, 33, 7)$

D) $(7, 33, -8) \sim (-8, 31, 11) \sim (11, 35, -2) \sim (-2, 33, 28) \sim$
$(28, 23, -7) \sim (-7, 33, 8) \sim (8, 31, -11) \sim (-11, 35, 2) \sim$
$(2, 33, -28) \sim (-28, 23, 7)$

A. O. L. Atkin has provided a labelling of the different kinds of cycles. Although the reasons for the existence or nonexistence of such cycles for a given discriminant will not appear until later, this labelling is now presented as an observation about examples.

Type 11: The complete ambiguous cycle is
$(1, a, -1) \sim (-1, a, 1)$.

Type 21: The complete ambiguous cycle is
$(a, abn, b) \sim (b, abn, a)$.

Type 12: The ambiguous cycle contains
$(a, ab, c) \sim \cdots \sim (x, y, -x) \sim \cdots \sim (-a, ab, -c) \sim \cdots$.

Type 22: The ambiguous cycle contains
$(a, ab, c) \sim \cdots \sim (f, de, d) \sim \cdots$
but does not contain the form $(-a, ab, -c)$.

Type 20: The ambiguous cycle contains
$(x, y, -x) \sim \cdots \sim (w, z, -w) \sim \cdots$.

Type 23: The cycle, which is not ambiguous, contains twice an even number of forms.

Type 13: The cycle, which is not ambiguous, contains twice an odd number of forms.

For reasons to be explained in the next section, the negative Pell equation $x^2 - \Delta y^2 = -4$ is solvable exactly for discriminants $\Delta$ which have cycles of Types 11, 12, and/or 13. For odd discriminants, the first occurrences of the different types are given below.

11) Type 11, $D = 5$, the cycle being
$(1, 1, -1) \sim (-1, 1, 1)$;

21) Type 21, $D = 21$, the cycle being
$(1, 3, -3) \sim (-3, 3, 1)$;

12) Type 12, $D = 17$, the cycle being
$(1, 3, -2) \sim (-2, 1, 2) \sim (2, 3, -1) \sim (-1, 3, 2) \sim$
$(2, 1, -2) \sim (-2, 3, 1)$;

22) Type 22, $D = 33$, the cycle being
$(1, 5, -2) \sim (-2, 3, 3) \sim (3, 3, -2) \sim (-2, 5, 1)$;

20) Type 20, $D = 205$, the cycle being
$(7, 3, -7) \sim (-7, 11, 3) \sim (3, 13, -3) \sim (-3, 11, 7)$;

23) Type 23, $D = 321$, the cycle being
$(5, 9, -12) \sim (-12, 15, 2) \sim (2, 17, -4) \sim (-4, 15, 6) \sim$
$(6, 9, -10) \sim (-10, 11, 5)$;

13) Type 13, $D = 145$, the cycle being
$(3, 7, -8) \sim (-8, 9, 2) \sim (2, 11, -3) \sim (-3, 7, 8) \sim$
$(8, 9, -2) \sim (-2, 11, 3)$.

**Examples**

| $\Delta$ | $h$ | Cycles |
|---|---|---|
| 8 | 1 | $(1,2,-1)(-1,2,1)$ |
| 12 | 2 | $(1,2,-2)(-2,2,1)$ |
| | | $(-1,2,2)(2,2,-1)$ |
| 20 | 1 | $(1,4,-1)(-1,4,1)$ |
| 24 | 2 | $(1,4,-2)(-2,4,1)$ |
| | | $(-1,4,2)(2,4,-1)$ |
| 28 | 2 | $(1,4,-3)(-3,2,2)(2,2,-3)(-3,4,1)$ |
| | | $(-1,4,3)(3,2,-2)(-2,2,3)(3,4,-1)$ |
| 32 | 2 | $(1,4,-4)(-4,4,1)$ |
| | | $(-1,4,4)(4,4,-1)$ |
| 40 | 2 | $(1,6,-1)(-1,6,1)$ |
| | | $(2,4,-3)(-3,2,3)(3,4,-2)(-2,4,3)(3,2,-3)(-3,4,2)$ |
| 44 | 2 | $(1,6,-2)(-2,6,1)$ |
| | | $(-1,6,2)(2,6,-1)$ |
| 48 | 2 | $(1,6,-3)(-3,6,1)$ |
| | | $(-1,6,3)(3,6,-1)$ |
| 52 | 1 | $(1,6,-4)(-4,2,3)(3,4,-3)(-3,2,4)(4,6,-1)$ |
| | | $(-1,6,4)(4,2,-3)(-3,4,3)(3,2,-4)(-4,6,1)$ |
| 5 | 1 | $(1,1,-1)(-1,1,1)$ |
| 13 | 1 | $(1,3,-1)(-1,3,1)$ |
| 17 | 1 | $(1,3,-2)(-2,1,2)(2,3,-1)(-1,3,2)(2,1,-2)(-2,3,1)$ |
| 21 | 2 | $(1,3,-3)(-3,3,1)$ |
| | | $(-1,3,3)(3,3,-1)$ |
| 29 | 1 | $(1,5,-1)(-1,5,1)$ |
| 33 | 2 | $(1,5,-2)(-2,3,3)(3,3,-2)(-2,5,1)$ |
| | | $(-1,5,2)(2,3,-3)(-3,3,2)(2,5,-1)$ |
| 37 | 1 | $(1,5,-3)(-3,1,3)(3,5,-1)(-1,5,3)(3,1,-3)(-3,5,1)$ |
| 41 | 1 | $(1,5,-4)(-4,3,2)(2,5,-2)(-2,3,4)(4,5,-1)$ |
| | | $(-1,5,4)(4,3,-2)(-2,5,2)(2,3,-4)(-4,5,1)$ |
| 45 | 2 | $(1,5,-5)(-5,5,1)$ |
| | | $(-1,5,5)(5,5,-1)$ |
| 53 | 1 | $(1,7,-1)(-1,7,1)$ |

## 3.2 Automorphs, Pell's Equation

The equation $x^2 - dy^2 = 1$, with $d$ a fixed integer and $x$ and $y$ assumed to be integer variables, has been called Pell's equation, although this is, in fact, a misattribution due to Euler. We shall, in general, refer to the equations $x^2 - \Delta y^2 = \pm 4$ as *Pell's equations*, and the equation with only the minus sign as the *negative Pell equation*; we note that if $\Delta$ is a discriminant of binary quadratic forms, then the existence of a solution to the Pell equations implies the existence of a solution to $x^2 - \Delta y^2 = \pm 1$, where the $\pm$ signs correspond. We recall that an automorph of a binary quadratic form is a nontrivial transformation (1.1) of determinant $+1$ under which the form is equivalent to itself.

**Theorem 3.9.** *If $\Delta$ is any discriminant of binary quadratic forms, then there exists a solution $(x, y)$ to the Pell equation*

$$x^2 - \Delta y^2 = 4. \tag{3.2}$$

*There is a one-to-one correspondence between automorphs of (definite or indefinite) forms $(a,\ b,\ c)$ of discriminant $\Delta$ and solutions of the Pell equation (3.2).*

**Proof.** We have defined the *principal root* of a form for positive definite forms; for indefinite forms the definition is identical:

$$\omega = \frac{-b + \sqrt{\Delta}}{2a}.$$

Now, if an automorph exists for a reduced form under a transformation (1.1), then

$$\omega = \frac{\alpha \omega + \beta}{\gamma \omega + \delta}.$$

This can be rewritten as a quadratic equation in $\omega$:

$$\gamma\omega^2 + \omega(\delta - \alpha) - \beta = 0.$$

But we already have $a\omega^2 + b\omega + c = 0$, and the form $(a,\ b,\ c)$ is assumed to be primitive, so we must have $\gamma = ka$, $\delta - \alpha = kb$, and $\beta = -kc$ for $k$ an integer. This gives

$$(\delta - \alpha)^2 + 4\gamma\beta = \Delta k^2.$$

We reduce this to get

$$(\alpha + \delta)^2 - \Delta k^2 = 4.$$

Given any automorph of a reduced form, then, we have a solution of the Pell equation $x^2 - \Delta y^2 = 4$. Given any solution of that equation, conversely, we have integers

$$
\begin{aligned}
\alpha &= (x - by)/2, \\
\beta &= -cy, \\
\gamma &= ay, \\
\delta &= (x + by)/2,
\end{aligned}
$$

and an automorph of the form $(a,\ b,\ c)$. The correspondence between automorphs and solutions is clear; only the existence of solutions is yet in question.

If $\Delta$ is a negative discriminant, then the equation is solvable only for $\Delta = -3$ or $-4$, and, of course, only for $+1$ on the right-hand side. This case was covered in Chapter 2.

It is only necessary, then, to consider positive discriminants $\Delta$. If we begin with the principal form of discriminant $\Delta$ and move through the principal cycle, we obtain transformation matrices which produce from a reduced form the equivalent adjacent reduced form. At some point we finish the cycle and return to the principal form. The product of all the transformation matrices is thus a transformation matrix which takes the principal form to itself. Since it cannot, except in trivial instances, be the identity matrix, it is the matrix of an automorph of the principal form. From this we get a solution to (3.2), and the theorem is proved. An example is given at the end of this chapter.

For the remainder of this chapter, only positive discriminants $\Delta = D$ are considered. Among all the solutions $(X, Y)$ to (3.2), there exists one for which $X$ and $Y$ are positive and $(X + Y\sqrt{D})/2$ is of least magnitude. We call this the *fundamental solution* of (3.2), noting that if $X'$ and $Y'$ are positive and $(X', Y')$ is another solution of (3.2), then $X < X'$ and $Y < Y'$ must also be true.

**Theorem 3.10.** *All pairs $(X_n, Y_n)$ generated by*

$$\frac{(X + Y\sqrt{D})^n}{2^n} = \frac{X_n + Y_n\sqrt{D}}{2}, \quad n \geq 1 \qquad (3.3)$$

*are solutions of equation (3.2). All solutions of equation (3.2) in positive rational integers are given by (3.3).*

**Proof.** That $X_n$ and $Y_n$ are rational integers follows by induction and observations about the parity of $X$, $Y$, and $D$. We then observe that

$$\frac{(X - Y\sqrt{D})^n}{2^n} = \frac{X_n - Y_n\sqrt{D}}{2} \qquad (3.4)$$

and thus

$$\frac{X_n^2 - DY_n^2}{4} = \frac{(X^2 - DY^2)^n}{4^n} = 1.$$

This proves the first part. To prove the second part, assume that another solution $(T, U)$ exists. Then there exists an $n \geq 1$ such that

$$\frac{(X + Y\sqrt{D})^n}{2^n} < \frac{T + U\sqrt{D}}{2} < \frac{(X + Y\sqrt{D})^{n+1}}{2^{n+1}}.$$

We multiply by the (positive) value $(X_n - Y_n\sqrt{D})/2$ and get

$$2 < T' + U'\sqrt{D} < X + Y\sqrt{D},$$

with $2T' = TX_n + UY_n$ and $2U' = TY_n + UX_n$. Again, by parity arguments, $T'$ and $U'$ are integral. Now, since $T' + U'\sqrt{D} > 2$ and $(T' + U'\sqrt{D}) \cdot (T' - U'\sqrt{D}) = 4$, we find $0 < T' - U'\sqrt{D} < 2$, which allows us to see that $T'$ and $U'$ are both positive. This, however, would contradict the fact that $(X + Y\sqrt{D})/2$ was the fundamental solution.

# 3.3 Continued Fractions and Indefinite Forms

We define a *continued fraction expansion* of $x$ (cf) to be a function

$$x = f(a_0, \ldots, a_N) = a_0 + \cfrac{1}{a_1 + \cfrac{\cdots}{+\frac{1}{a_N}}} \tag{3.5}$$

At present $x$ may be any sort of number, although soon only rational numbers and real quadratic irrationals will be considered. We define the values $a_i$ to be the *partial quotients* of the cf. The above cf will be abbreviated as

$$[a_0, \ldots, a_N],$$

whose *n-th convergent* is

$$R(n) = [a_0, \ldots, a_n], \quad 0 \leq n \leq N.$$

**Theorem 3.11.** *Defining*

$$\begin{aligned} P_{-1} &= 1 \\ P_0 &= a_0 \\ P_n &= a_n \cdot P_{n-1} + P_{n-2}, \; for \; n \geq 1, \end{aligned}$$

*and*

$$\begin{aligned} Q_{-1} &= 0 \\ Q_0 &= 1 \\ Q_n &= a_n \cdot Q_{n-1} + Q_{n-2}, \; for \; n \geq 1, \end{aligned}$$

*then*

$$R_n = [a_0, \ldots, a_n] = P_n/Q_n \ for \ n \geq 0. \tag{3.6}$$

**Proof.** The theorem holds for $n = 0$. We assume that it holds for $n \leq m$ and calculate

$$
\begin{aligned}
[a_0, \ldots, a_{m+1}] &= [a_0, \ldots, a_m + 1/a_{m+1}] \\
&= \frac{(a_m + 1/a_{m+1}) \cdot P_{m-1} + P_{m-2}}{(a_m + 1/a_{m+1}) \cdot Q_{m-1} + Q_{m-2}} \\
&= \frac{a_{m+1} \cdot (a_m P_{m-1} + P_{m-2}) + P_{m-1}}{a_{m+1} \cdot (a_m Q_{m-1} + Q_{m-2}) + Q_{m-1}} \\
&= \frac{a_{m+1} P_m + P_{m-1}}{a_{m+1} Q_m + Q_{m-1}} \\
&= \frac{P_{m+1}}{Q_{m+1}},
\end{aligned}
$$

where the penultimate equality is by induction.

There are three other formulas of interest, all of which can be proved by direct calculation and/or recursion:

$$P_n Q_{n-1} - P_{n-1} Q_n = (-1)^{n-1}, \ n \geq 0, \tag{3.7}$$

$$\frac{P_n}{Q_n} - \frac{P_{n-1}}{Q_{n-1}} = \frac{(-1)^{n-1}}{Q_n Q_{n-1}}, \ n \geq 0, \tag{3.8}$$

$$P_n Q_{n-2} - P_{n-2} Q_n = a_n(-1)^{n-1}, \ n \geq 1. \tag{3.9}$$

We now restrict ourselves to the case when each $a_i$, $i \geq 1$, is positive and integral. These are called *simple* continued fractions (scf's). In this case any finite scf represents a rational number $x$. Before proving the converse, we define the

*Continued Fraction Algorithm:* Define $a_i$, $X_i$, and $Z_i$ by

$$x = a_0 + Z_0, \text{chosen so that } 0 \le Z_0 < 1, \qquad (3.10)$$
$$X_i = 1/Z_{i-1} = a_i + Z_i, \ i \ge 1, \ 0 \le Z_i < 1.$$

The algorithm continues as long as $Z_i \ne 0$. The (not necessarily integral) values $X_i$ are the *i-th complete quotients* in the cf expansion, that is,

$$x = [a_0, \ldots, X_i].$$

**Theorem 3.12.** *Any rational number $x$ has a representation as a finite simple continued fraction.*

**Proof.** We shall not prove this. The proof is straightforward; indeed it is a rephrasing of the usual algorithm for computing the greatest common divisor of the numerator and denominator of the rational number $x$.

Example: Let $x = 267/111$. Then computing in order $a_0$ , $Z_0$ , $X_1$ , $a_1$ , $Z_1$ , ..., and then $P_i$ and $Q_i$ afterwards, we have

| $i$ | $a_i$ | $P_i$ | $Q_i$ | $Z_i$ | $X_i$ | $\mid P_i \cdot 111 - Q_i \cdot 267 \mid$ |
|-----|-------|-------|-------|-------|-------|-------------------------------------------|
| $-1$ |      | 1     | 0     |       |       | 111 |
| 0   | 2     | 2     | 1     | 45/111 |      | 45 |
| 1   | 2     | 5     | 2     | 21/45 | 111/45 | 21 |
| 2   | 2     | 12    | 5     | 3/21  | 45/21 | 3 |
| 3   | 7     | 89    | 37    | 0     | 21/3  | 0 |

We now have one more major list of facts.

**Theorem 3.13.** *Let* $x = [a_0, \ldots, a_N]$ *be a finite scf. Then*

a) $R_{2n} < R_{2n+2}$ *and* $R_{2n-1} > R_{2n+1}$ *for all* $n \geq 0$.

b) $R_{2n} < R_{2i+1}$ *for all* $n, i \geq 0$.

c) $R_{2n} < x$ *and* $R_{2n-1} > x$ *for all convergents except the last.*

d) $Q_n > Q_{n-1}$ *for all* $n > 1$

e) $\gcd(P_n, Q_n) = 1$ *for all* $n$.

**Proof.** *a)* Looking at (3.9), and remembering that the $a_i$ and $Q_i$ are all positive, we see that the right-hand side of (3.9) is positive or negative according as $n$ is even or odd.

*b)* In (3.8), it is clear that $R_{2n} < R_{2n+1}$ for all $n$. If $R_{2n} > R_{2i+1}$ were to hold for some $n < i$, then $R_{2i} > R_{2i+1}$ would hold since by part a $R_{2n}$ is an increasing sequence. Similarly, if $R_{2n} > R_{2i+1}$ were to hold for some $n > i$, then $R_{2n} > R_{2n+1}$ would hold since the $R_{2n+1}$ are decreasing. These are both contradictions.

*c)* This is obvious. $x$ has some value, which is larger than the even convergents and smaller than the odd ones, except for the equality which holds for the last.

*d)* This is evident from the defining equations of Theorem 3.12 and the new assumption that the $a_i$ are positive.

*e)* In (3.7), the gcd of $P_n$ and $Q_n$ must divide either $-1$ or $+1$ and hence must be 1.

We now pass from finite scf's to infinite ones.

**Theorem 3.14.** *If* $a_0$ *is an integer and* $a_1, \ldots, a_n, \ldots$ *is any sequence* *of positive integers, then*

$$x = \lim_{n \to \infty} [a_0, \ldots, a_n, \ldots]$$

*exists, and is greater than any even convergent and smaller than any*

*odd convergent.*

**Proof.** The even convergents are increasing and the odd convergents are decreasing, so if the limit exists the rest must be true. But by (3.8) and Theorem 3.13d we have

$$\mid R_{n+1} - R_n \mid = \frac{1}{Q_n Q_{n+1}} < \frac{1}{Q_{n+1}^2}.$$

The right hand side goes to 0 as $n \to 0$, so the limit does exist.

From this point on, only *periodic* simple continued fractions are considered, that is, scf's for which $a_i = a_{i+J}$ for all $i \geq I$ and some fixed $J$. We write this as

$$[a_0, \ldots, a_{I-1}, *a_I, \ldots, *a_{I+J-1}],$$

with the * indicating the period. We can now, at last, prove one of the main theorems and return to the discussion of quadratic forms.

**Theorem 3.15.**

  a) *If $\omega$ is an irrational root of a quadratic equation with integer coefficients, then the scf for $\omega$ is periodic.*

  b) *If an scf is periodic, then its value is an irrational root of a quadratic equation with integer coefficients.*

**Proof.** *a)* Let $\omega$ be the root of

$$a\omega^2 + b\omega + c = 0,$$

where without loss of generality we have $a > 0$. Writing $D = b^2 - 4ac$,
we can see that $Z_1 = (-B + \sqrt{D})/(2A)$ with $A > 0$, $0 < B < \sqrt{D}$,
and $B^2 - D = 4AC$ for a positive integer C. From there it is clear
that all of the $Z_i$ are of this form. But this limits the values of $B$ to
a finite list, and consequently there are only finitely many values $Z_i$
which occur. Clearly, then, the cf is periodic since the choice of $Z_{i+1}$
from $Z_i$ is unique.

b) Let $\omega = [a_0, \ldots, a_{I-1}, *a_I, \ldots, *a_{I+J-1}]$. Then, in the notation of
(3.10), $X_I$ is the value of the purely periodic part. If $P'/Q'$ and $P''/Q''$
are the last two convergents of $[a_I, \ldots, a_{I+J-1}]$, then

$$X_I = [*a_I, \ldots, *a_{I+J-1}, X_I]$$

so that

$$X_I = \frac{P'X_I + P''}{Q'X_I + Q''}.$$

(The left-hand $X_I$ is the value of the periodic scf; the right-hand $X_I$
are from applying the defining recursions of the $P$ and $Q$ convergents.)
Thus, $X_I$ is a quadratic irrational, satisfying an equation with integer
coefficients. Now, we can also write

$$x = \frac{P_{I-1}X_I + P_{I-2}}{Q_{I-1}X_I + Q_{I-2}},$$

and hence $x$ is a quadratic irrational, satisfying an equation which can
be seen to have integer coefficients.

We now return to our main topic. Given a discriminant of binary
quadratic forms $D > 0$, we define $\omega = \sqrt{D/4}$, if $D$ is even, and $\omega = (-1 + \sqrt{D})/2$, if $D$ is odd. These are the principal roots of forms
$(1, 0, -D/4)$ and $(1, 1, (1-D)/4)$, respectively, which forms we write

as $f = (1, b, (b-D)/4)$. Expanding the cf for $\omega$ produces $\omega = a_0 + Z_0$, with $0 \le Z_0 < 1$ and $a_0$ an integer. Under the transformation

$$\begin{pmatrix} 1 & a_0 \\ 0 & 1 \end{pmatrix},$$

$f$ becomes $(1, b+2a_0, *)$, where $b+2a_0 < \sqrt{D} < b+2a_0+2$. This form is thus reduced, with principal root $Z_0$. At this point, expansion of the cf and cycling through the principal cycle of forms of discriminant $D$ are essentially the same.

Let us now consider the equivalences

$$(1, b, (b-D)/4) \sim (1 = c_0, b_0, c_1) \sim (c_1, b_1, c_2) \sim \ldots$$

under the action of the above transformations $T_i$. Then the sequences $Z_i = (-b_i + \sqrt{D})/(2c_i)$ and $X_i = (b_i + \sqrt{D})/(2c_{i+1})$ are clear. We have the following theorem.

**Theorem 3.16.** *If $M = T_0 \ldots T_i$ transforms $(1, b, (b-D)/4)$ into $(c_i, b_i, c_{i+1})$, then*

$$(2P_i + Q_i)^2 - DQ_i^2 = 4c_{i+1}.$$

**Proof.** The proof follows from writing

$$\omega = \frac{X_{n+1}P_n + P_{n-1}}{X_{n+1}Q_n + Q_{n-1}}.$$

The rest follows simply by calculation.

In the expansion of the cf it may happen that $(1, b, (D-b)/4)$ and $(-1, b, (b-D)/4)$ lie in the same cycle; if this is true, the cycle of forms is twice as long as the period of the cf, with the cf cycle being repeated in the period of forms. If this happens, we choose to call the length of the cf period to be the same as the length as the cycle of forms. (This is the case in our example at the end of this chapter. The cf for $(-5 + \sqrt{41})/2$ has a period of length 10 and not 5.) With this convention, we obtain two theorems which together give us the precise determination of the solutions to the Pell equations.

**Theorem 3.17.** *If the continued fraction expansion of $\omega = (-1 + \sqrt{D})/2$ (for odd D) or of $\omega = \sqrt{D}/2$ (for even D) is of length $n$, and if $P = P_{n-1}$ and $Q = Q_{n-1}$ are the penultimate convergents in the first period of the expansion, then $(X, Y) = (2P + Q, Q)$ is the fundamental solution of (3.2).*

**Proof.** Clearly $(2P+Q, Q)$ is a solution, but then $(2P+Q+Q\sqrt{D})/2 = ((X + Y\sqrt{D})/2)^n$ for some $n$. Then the expansion of the cf for $(2P + Q + Q\sqrt{D})/2$ contains $n$ copies of the cf for the fundamental solution of (3.2). But no such repetition can occur, except for the double period that occurs if, as mentioned above, the $(-1, *, **)$ form appears in the principal cycle; however, that would, by Theorem 3.16, provide a solution to $x^2 - Dy^2 = -4$.

The following theorem can now be proved by carefully combining previous results.

**Theorem 3.18.** *Let $\Delta = D$ be a positive discriminant of quadratic forms. Solutions to the Pell equation*

$$x^2 - Dy^2 = -4 \qquad\qquad (3.11)$$

*exist if and only if the reduced forms $(1, b, c)$ and $(-1, b, -c)$ of discriminant $D$ lie in the same cycle. If this is true, then*

a) *the length of the continued fraction expansion of* $\omega = (-1+\sqrt{D})/2$
   *(for odd D) or of* $\omega = \sqrt{D}/2$ *(for even D) (which is the length of
   the cycle of forms) is an even integer 2n;*

b) *if* $P = P_{n-1}$ *and* $Q = Q_{n-1}$ *are the penultimate convergents in
   the first half-period of the expansion, then* $(X,Y) = (2P+Q, Q)$
   *is the solution of (3.11) for which X and Y are positive integers
   and* $(X + Y\sqrt{D})/2$ *is of least magnitude;*

c) *all solutions to (3.11) are given by the odd powers of* $(X+Y\sqrt{D})/2$;

d) *all solutions to (3.2) are given by the even powers of* $(X+Y\sqrt{D})/2$;
   *the fundamental solution to (3.2) is*

$$\left(\frac{X+Y\sqrt{D}}{2}\right)^2.$$

The solution to (3.11), if it exists, will be called the *fundamental solution* to that equation.

We now prove Theorem 3.5.

**Theorem 3.5.** *Two reduced forms are equivalent if and only if they
are in the same cycle.*

**Proof.** Our proof, which follows closely that of Mathews, will take
several steps. We define a continued fraction to be *regular* if all the
partial quotients after the first are positive.

**Proposition 3.19.** *If an infinite cf contains only a finite number of
nonpositive partial quotients, it can be converted in a finite number of
steps to a regular cf.*

**Proof.** Let $a_r$ be the last nonpositive partial quotient (pq).

*Case i.* $a_r = 0$.

Since $[x, 0, y, z] = [x + y, z]$, we have

$$[\ldots, a_{r-1}, 0, a_{r+1}, a_{r+2}, \ldots] = [\ldots, a_{r-1} + a_{r+1}, a_{r+2}, \ldots].$$

We note that this shifts the last nonpositive pq to the left.

*Case ii.* $a_r = -k \neq -1$.

It can be shown that

$$[\ldots, a_{r-1}, -k, a_{r+1}, \ldots] = [\ldots, a_{r-1} - 1, k - 2, 1, a_{r-1} - 1, \ldots].$$

Since $a_r$ is the last nonpositive pq, $a_{r+1} - 1$ is nonnegative. If it is zero

or if $k$ is 2, the reduction of the previous case has the effect of shifting

the last nonpositive pq to the left.

*Case iii.* $a_r = -1$.

Since $[\ldots, x, -1, y, \ldots] = [\ldots, x - 2, 1, y - 2, \ldots]$

and $[\ldots, x, -1, 1, y, z, \ldots] = [\ldots, x - y - 2, 1, z - 1, \ldots]$,

the last nonpositive pq is again shifted to the left.

We can thus shift the nonpositive terms to the left, eventually eliminating them entirely. In each case the number of partial quotients
changes by zero or by two.

**Proposition 3.20.** *If $y = (\alpha x + \beta)/(\gamma x + \delta)$ for some transformation
in the modular group $\Gamma$, then $y$ can be written*

$$y = [\pm t, a_1, \ldots, a_{2r}, \pm u, x] \, ,$$

*with $a_1, \ldots, a_{2r}$ all positive.*

**Proof.** Let $\pm t$ be chosen so that $-(\pm t - \beta/\delta)$ is a positive proper
fraction. We can expand $\beta/\delta$ into a cf with an odd number of partial

quotients $\beta/\delta = [\pm t, a_1, \ldots, a_{2r}]$. (We can make the length $2r+1$ since $[z] = [z-1, 1]$.) If $P/Q$ is the penultimate convergent, then by (3.7), $\beta Q - \delta P = 1 = \alpha\delta - \beta\gamma$. Then $\alpha = P \pm u\beta$ and $\gamma = Q \pm u\delta$ for some integer $u$. Then

$$\alpha/\gamma = [\pm t, a_1, \ldots, a_{2r}, \pm u].$$

Consequently,

$$y = \frac{\alpha x + \beta}{\gamma x + \delta} = [\pm t, a_1, \ldots, a_{2r}, \pm u, x].$$

We now prove Theorem 3.5. Let $f = (a, \ b, \ c)$ and $f' = (a', \ b', \ c')$ be two reduced equivalent forms. With no loss of generality for our purposes, we can choose $a$ and $a'$ positive so that the principal roots $\omega$ and $\omega'$ are positive proper fractions. Since the forms are equivalent, a transformation of the usual sort exists so that

$$\omega' = (\alpha\omega + \beta)/(\gamma\omega + \delta).$$

Then $\omega' = [\pm t, a_1, \ldots, a_{2r}, \pm u, \omega]$

$$= [\pm t, a_1, \ldots, a_{2r}, \pm u + d_1, *d_2, \ldots, *d_1]$$

if $[*d_1, \ldots, *d_{2m}]$ is the cf for $\omega$. We may use Proposition 3.18 to make all the partial quotients after the first positive and then note that the first partial quotient is zero since $\omega'$ is a positive proper fraction. It is easy to show that a purely periodic cf is unique for a given quadratic irrational so that the periodic part of the expansion of $\omega'$ is merely a cyclic permutation of that for $\omega$. Indeed, since the operations of Proposition 3.18 change the number of partial quotients by zero or two each time, the period for $\omega'$ is a shift of that for $\omega$ by an even number of partial quotients. (This is important since the first coefficients of adjacent forms alternate in sign; without the evenness of the permutation we could not distinguish cycles from their associates.) Thus, by cycling forward from $f$ we arrive at a reduced form whose principal root is $\omega'$. But the principal root and the discriminant uniquely determine the form, so this form is $f'$.

We prove one final theorem which will be used later.

**Theorem 3.21.** *Let $\Delta$ be a positive discriminant of binary quadratic forms and $p$ be any prime. In the notation of Theorem 3.9, we have that*

a) *there exists an $n$ such that $p \mid Y_n$;*

b) *the least positive residues modulo $p$ of the integers $(X_n, Y_n)$ form a periodic sequence.*

**Proof.** We only prove this for odd primes $p$; the proof for $p = 2$ is similar. If $\Delta$ is a discriminant of forms, then so is $\Delta p^2$; therefore, a solution exists to the equation $x^2 - \Delta p^2 y^2 = 4$. Part a follows from the fact that this solution $(x, py)$ to $x^2 - \Delta y^2 = 4$ must be one of the pairs $(X_n, Y_n)$.

We have that

$$
\begin{aligned}
\frac{X_p + Y_p\sqrt{\Delta}}{2} &= \left(\frac{X_1 + Y_1\sqrt{\Delta}}{2}\right)^p \\
&\equiv \frac{X_1 + Y_1\Delta^{(p-1)/2}\sqrt{\Delta}}{2} \\
&\equiv \frac{X_1 + Y_1\left(\frac{\Delta}{p}\right)\sqrt{\Delta}}{2},
\end{aligned}
$$

where the congruences are taken modulo $p$, and the symbol $\left(\frac{\Delta}{p}\right)$ is the quadratic residue symbol if $p$ does not divide $\Delta$ and 0 if it does. We may thus define $\bar{X} \equiv X_1$ and $\bar{Y} \equiv \left(\frac{\Delta}{p}\right)Y_1$ to be least positive residues of these congruences (mod $p$). It is clear that the powers $(X_1 + Y_1\sqrt{\Delta})/2)^n$ produce a sequence congruent modulo $p$ to the powers of $(\bar{X} + \bar{Y}\sqrt{\Delta})/2)^n$, and this sequence is recurrent.

*Example.*

Let $D = 41$. The cf expansion of $(-1 + \sqrt{41})/2$ is

| $i$ | $a_i$ | $P_i$ | $Q_i$ | $Z_i$ | $X_i$ |
|---|---|---|---|---|---|
| $-1$ | | 1 | 0 | | |
| 0 | 2 | 2 | 1 | $(\sqrt{41} - 5)/2$ | |
| 1 | 1 | 3 | 1 | $(\sqrt{41} - 3)/8$ | $(\sqrt{41} + 5)/8$ |
| 2 | 2 | 8 | 3 | $(\sqrt{41} - 5)/4$ | $(\sqrt{41} + 3)/4$ |
| 3 | 2 | 19 | 7 | $(\sqrt{41} - 3)/4$ | $(\sqrt{41} + 5)/4$ |
| 4 | 1 | 27 | 10 | $(\sqrt{41} - 5)/8$ | $(\sqrt{41} + 3)/8$ |
| 5 | 5 | 154 | 57 | $(\sqrt{41} - 5)/2$ | $(\sqrt{41} + 5)/2$ |
| 6 | 1 | 181 | 67 | $(\sqrt{41} - 3)/8$ | $(\sqrt{41} + 5)/8$ |
| 7 | 2 | 516 | 191 | $(\sqrt{41} - 5)/4$ | $(\sqrt{41} + 3)/4$ |
| 8 | 2 | 1213 | 449 | $(\sqrt{41} - 3)/4$ | $(\sqrt{41} + 5)/4$ |
| 9 | 1 | 1729 | 640 | $(\sqrt{41} - 5)/8$ | $(\sqrt{41} + 3)/8$ |
| 10 | 5 | 9858 | 3649 | $(\sqrt{41} - 5)/2$ | $(\sqrt{41} + 5)/2$ |

The cycle is completed, and the cf is [2, *1, 2, 2, 1, *5]. The effect on the forms is this, where the equivalences after the first are done with transformations

$$T_i = \begin{pmatrix} 0 & -1 \\ 1 & \delta \end{pmatrix},$$

for which the $\delta$ are $-1, 2, -2, 1, -5, 1, -2, 2, -1$:

$(1, 1, -10) \sim (1, 5, -4) \sim (-4, 3, 2) \sim (2, 5, -2) \sim$
$\qquad (-2, 3, 4) \sim (4, 5, -1) \sim (-1, 5, 4) \sim (4, 3, -2) \sim$
$\qquad (-2, 5, 2) \sim (2, 3, -4) \sim (-4, 5, 1).$

The cumulative equivalence is achieved by the transformations computed as follows:

$$T_0 = \begin{pmatrix} P_{-1} & P_0 \\ Q_{-1} & P_0 \end{pmatrix}$$

$$T_0T_1 \; = \; T_0 \begin{pmatrix} 0 & -1 \\ 1 & -a_1 \end{pmatrix}$$

$$= \begin{pmatrix} P_0 & -P_1 \\ Q_0 & -Q_1 \end{pmatrix}$$

$$T_0T_1T_2 \; = \; \begin{pmatrix} -P_1 & -P_2 \\ -Q_1 & -Q_2 \end{pmatrix}$$

$$T_0T_1T_2T_3 \; = \; \begin{pmatrix} -P_2 & P_3 \\ -Q_2 & Q_3 \end{pmatrix}$$

$$T_0T_1T_2T_3T_4 \; = \; \begin{pmatrix} P_3 & P_4 \\ Q_3 & Q_4 \end{pmatrix}.$$

Thus, for example, $(1,1,-10) \sim (2,5,-2)$ under

$$\begin{pmatrix} -3 & -8 \\ -1 & -3 \end{pmatrix}.$$

We see that $(1,5,-4) \sim (-1,5,4)$ under

$$\begin{pmatrix} 7 & -40 \\ 10 & -57 \end{pmatrix}.$$

This provides us with the solution $64^2 - 41 \cdot 10^2 = -4$, which is the fundamental solution for (3.11). Continuing to the end of the cycle, we find that $(1,5,-4)$ first becomes equivalent to itself under

$$\begin{pmatrix} -449 & -2560 \\ -640 & -3649 \end{pmatrix}.$$

From this we get the solution $4098^2 - 41 \cdot 640^2 = 4$, which is the fundamental solution for (3.2).

# Chapter 4

# The Class Group

## 4.1 Representation and Genera

We come again to the subject of the representation of integers by forms. Let us assume that, for a form $f = (a, b, c)$ of discriminant $\Delta$, integers $x$ and $y$ exist so that $f$ represents $r$, that is, $r = ax^2 + bxy + cy^2$. This is a *primitive representation* if $\gcd(x, y) = 1$. If the representation is primitive, then integers $z$ and $w$ exist so that $xw - yz = 1$. Then $f$ is equivalent to a form $f' = (r, s, t)$, where $f'$ is obtained from $f$ by using the transformation

$$\begin{pmatrix} x & y \\ z & w \end{pmatrix},$$

and equations (1.2). We note that the choice of $f'$ is not unique, but that different values of $s$ differ by multiples of $2r$, and thus the different choices lead to equivalent forms. That is, modulo $2r$, a unique $s$ is determined from $(x, y)$ such that

$$s^2 - 4rt = \Delta, \quad \text{for some integer } t.$$

**Proposition 4.1.** *If $f = (a, b, c)$ of discriminant $\Delta$ represents $r$, then*

$$s^2 - \Delta \equiv 0 \quad (\text{mod } 4r) \qquad\qquad (4.1)$$

*has an integer solution $s$. Conversely, if $\Delta$ is a discriminant of binary quadratic forms, and a solution $s$ exists to (4.1), then $r$ is represented by some form of discriminant $\Delta$.*

**Proof.** We have essentially proved this proposition. We note only that the existence of solutions to (4.1) is a stronger criterion than simply having the Jacobi symbol $\left(\frac{\Delta}{4r}\right)$ be $+1$ and that for odd $r$ the 4 in the modulus in (4.1) may be dropped.

By the above arguments, we also see that if $r$ is represented by a form $f$, then it is represented by all forms in the equivalence class of $f$. Indeed, any integers $x$ and $y$ used in the representation of $r$ by $f$ are directly linked to the transformation that takes $f$ to an equivalent form of lead coefficient $r$. Representation, then, is a property of classes, not of individual forms. We can go somewhat further than this, but first need a simplifying proposition. We have always dealt with primitive forms, but emphasize this at this point.

**Proposition 4.2.** *A primitive form can primitively represent an integer which is relatively prime to any chosen number.*

**Proof.** Consider the form $f = (a, b, c)$ and any integer $m$. Let $P$ be the product of the primes dividing $a$, $c$, and $m$, $Q$ the product of the primes dividing $a$ and $m$ but not $c$, $R$ the product of the primes dividing $c$ and $m$ but not $a$, and $S$ the product of the remaining primes dividing $m$. The form $f$ then represents $aQ^2 + bQRS + c(RS)^2$, which is relatively prime to $m$.

Now let us consider an odd prime $p$ that divides $\Delta$, and assume that a form represents $r$ and $r'$, both relatively prime to $p$:

$$r = ax^2 + bxy + cy^2$$
$$r' = ax'^2 + bx'y' + cy'^2.$$

Then $4rr' = A^2 - \Delta B^2$, where

$$A = 2axx' + b(xy' + x'y) + 2cyy'$$

and

$$B = xy' - x'y.$$

Then the Jacobi symbol $\left(\frac{rr'}{p}\right)$ is $+1$, so that $\left(\frac{r}{p}\right) = \left(\frac{r'}{p}\right)$. Thus, all representable integers prime to $p$ have the same quadratic character modulo $p$.

Now let a form $f$ represent odd numbers $r$ and $r'$. It is easy to see that for even discriminants, if $\Delta = 4\Delta'$, then

1) if $\Delta' \equiv 0 \pmod 8$, then $rr' \equiv 1 \pmod 8$;

2) if $\Delta' \equiv 1 \pmod 8$, then $rr' \equiv 1,3,5,7 \pmod 8$;

3) if $\Delta' \equiv 2 \pmod 8$, then $rr' \equiv 1,7 \pmod 8$;

4) if $\Delta' \equiv 3 \pmod 8$, then $rr' \equiv 1,5 \pmod 8$;

5) if $\Delta' \equiv 4 \pmod 8$, then $rr' \equiv 1,5 \pmod 8$;

6) if $\Delta' \equiv 5 \pmod 8$, then $rr' \equiv 1,3,5,7 \pmod 8$;

7) if $\Delta' \equiv 6 \pmod 8$, then $rr' \equiv 1,3 \pmod 8$;

8) if $\Delta' \equiv 7 \pmod 8$, then $rr' \equiv 1,5 \pmod 8$;

In cases 2 and 6, no information is gained; these, however, are cases when $\Delta'$ is also a discriminant of forms. The other cases can be grouped as follows:

1) if $\Delta' \equiv 0 \pmod 8$, then $\left(\frac{-1}{rr'}\right) = \left(\frac{2}{rr'}\right) = +1$;

2) if $\Delta' \equiv 2 \pmod 8$, then $\left(\frac{2}{rr'}\right) = +1$;

3) if $\Delta' \equiv 3, 4, 7 \pmod 8$, then $\left(\frac{-1}{rr'}\right) = +1$;

4) if $\Delta' \equiv 6 \pmod 8$, then $\left(\frac{2}{rr'}\right) = +1$;

We define the quadratic characters $\chi(r) = \left(\frac{-1}{r}\right)$ and $\psi(r) = \left(\frac{2}{r}\right)$.

The *generic characters* of a discriminant $\Delta$ are

0) $\left(\frac{r}{p}\right)$ for all odd primes $p$ that divide $\Delta$;

1) if $\Delta$ is even and $\Delta/4 \equiv 3, 4, 7 \pmod 8$, $\chi(r)$;

2) if $\Delta$ is even and $\Delta/4 \equiv 2 \pmod 8$, $\psi(r)$;

3) if $\Delta$ is even and $\Delta/4 \equiv 6 \pmod 8$, $\chi(r) \cdot \psi(r)$;

4) if $\Delta$ is even and $\Delta/4 \equiv 0 \pmod 8$, both $\chi(r)$ and $\psi(r)$;

We shall refer to each quadratic character $\left(\frac{r}{p}\right)$, $\chi$, $\psi$, or $\chi\psi$, as a generic character for a discriminant $\Delta$. The characters are multiplicative functions from the integers to $\{+1, -1\}$; we shall refer to the values $+1$ or $-1$ assigned by these functions to an integer $r$ as the *assigned values*. With some arbitrary but fixed ordering given to the set of $n$ characters, they can be considered as a vector-valued function from the integers to an $n$-tuple with entries either $+1$ or $-1$. The following proposition, that the vector of assigned values is invariant over all numbers representable by any form in a class, has been proved by the preceding arguments.

**Proposition 4.3.** *All the integers r relatively prime to* $\Delta$ *which are representable by forms in a given equivalence class possess the same assigned values of generic characters.*

The classes of forms possessing the same assigned values of generic characters are called a *genus* of forms. The genus which has all assigned characters $+1$ must contain the principal form and is called the *principal genus.* As the following example shows, not all possible vectors of assigned values need exist; we shall return to this issue in Section 4.3 and show that exactly half exist.

*Example.* $\Delta = -1520 = -16 \cdot 5 \cdot 19.$
The 16 reduced forms fall into the following genera:

| $\left(\frac{r}{5}\right)$ | $\left(\frac{r}{19}\right)$ | $\chi$ | |
|---|---|---|---|
| $+$ | $+$ | $+$ | $(1,0,380),(5,0,76),(9,\pm 8,44)$ |
| $+$ | $+$ | $-$ | $(4,0,95),(19,0,20),(11,\pm 8,36)$ |
| $-$ | $-$ | $+$ | $(12,\pm 8,33),(13,\pm 12,32)$ |
| $-$ | $-$ | $-$ | $(3,\pm 2,127),(15,\pm 10,27)$ |

We summarize the results by specifying exactly the representation of integers by forms.

**Theorem 4.4.** *An integer r is representable by some class of forms of discriminant* $\Delta$ *if and only if the assigned values of the generic characters of r match the assigned values of characters of some genus of discriminant* $\Delta$. *This is true if and only if the congruence*

$$s^2 = \Delta \quad (\mathrm{mod}\ 4r)$$

*is solvable. If* $\{s_i : i = 1, \ldots, k\}$ *are all the solutions of the congruence, then the classes of the forms* $\{(r, \pm s_i, (s_i{}^2 - \Delta)/4r) : i = 1, \ldots, k\}$ *are* *precisely the classes which represent $r$. If the transformation*

$$\begin{pmatrix} \alpha & \beta \\ \gamma & \delta \end{pmatrix}$$

*takes the form $(r, s_i, (s_i{}^2 - \Delta)/4r)$ to the form $(a, b, c)$, then $r = a\delta^2 - b\beta\delta + c\beta^2$ is one representation of $r$ by $(a, b, c)$, and all representations of $r$ are given by $r = ax^2 + bxy + cy^2$ with*

$$(x, y) = (T(2a\delta - b\beta) + U\beta\Delta, \pm[T\beta + U(2a\delta - b\beta)])$$

*or*

$$(x, y) = (T(2a\delta - b\beta) - U\beta\Delta, \pm[T\beta - U(2a\delta - b\beta)]),$$

*where $(T, U)$ is any solution to $T^2 - \Delta U^2 = 4$.*

## 4.2 Composition Algorithms

We now prove the last of the elementary results, that the classes of forms of a fixed discriminant form a finite abelian group (the *class group*). The existence and nature of the group of classes of forms was first discovered by Gauss and published in the *Disquisitiones Arithmeticae*.

In its historical context the rules of composition were first discovered as a generalization of a multiplication-of-forms formula known to ancient mathematicians:

$$(x^2 + Dy^2) \cdot (z^2 + Dw^2) = (xz + Dyw)^2 + D(xw - yz)^2.$$

The identity in the class group is the principal form $(1, \ 0, \ D)$ or $(1, \ 1, \ (1 - D)/4)$. The above formula in a class group context is the statement that the square of the identity of the group is the identity. The machinery of composition of forms is entirely for the purpose of extending this identity to its most general form, for all classes of forms and all possible values of $D$, which need not be the same in each of the two factors on the left-hand side.

The existence of a group structure is best seen by defining the form composition using the "united forms" of Dirichlet. Two forms $(a_1, \ b_1, \ c_1)$ and $(a_2, \ b_2, \ c_2)$ of the same discriminant $\Delta$ are called *united* if the greatest common divisor of $a_1$, $a_2$, and $(b_1 + b_2)/2$ is 1.

**Proposition 4.5.** *If* $(a_1, \ b_1, \ c_1)$ *and* $(a_2, \ b_2, \ c_2)$ *are united forms, then there exist forms* $(a_1, \ B, \ a_2C)$ *and* $(a_2, \ B, \ a_1C)$ *such that*

$$(a_1, \ b_1, \ c_1) \sim (a_1, \ B, \ a_2C)$$
$$(a_2, \ b_2, \ c_2) \sim (a_2, \ B, \ a_1C)$$

**Proof.** We must demonstrate that an integer $B$ exists such that the congruences

$$B \equiv b_1 \pmod{2a_1}$$
$$B \equiv b_2 \pmod{2a_2}$$

are simultaneously solvable and for which $C$ is an integer. The solutions of the first congruence are $b_1 + 2a_1\delta_1$. These provide a solution of the second congruence if and only if

$$\frac{b_1 - b_2}{2} \equiv -a_1\delta_1 \pmod{a_2}$$

is solvable, and this is true if and only if the greatest common divisor $d$ of $a_1$ and $a_2$ divides $(b_1 - b_2)/2$. From the equation

$$\Delta = b_1^2 - 4a_1c_1 = b_2^2 - 4a_2c_2$$

we may deduce that

$$\left(\frac{b_1 + b_2}{2}\right)\left(\frac{b_1 - b_2}{2}\right) = a_1c_1 - a_2c_2.$$

Since the forms are united, $d$ must divide $(b_1 - b_2)/2$.

Now let $k = a_1a_2/d$. The value of $B$ determined above is unique modulo $2k$, that is, $B = B_0 + 2kt$ for some fixed $B_0$ and arbitrary $t$. By our choice of $B_0$, we have $B^2 \equiv B_0^2 \equiv \Delta \pmod{4k}$. We now need to choose $t$ to force $B^2 \equiv \Delta \pmod{4a_1a_2}$; this will guarantee that $C$ is an integer.

Clearly, $B^2 \equiv \Delta \pmod{4a_1a_2}$ if and only if

$$\frac{\Delta - B_0^2}{4k} \equiv B_0 t \pmod{d}$$

and the left hand side is indeed an integer. By the definition of $B_0$ and the fact that the forms are united, $B_0$ is invertible modulo $d$, so

$$t \equiv \left(\frac{\Delta - B_j^2}{4k}\right) B_0^{-1} \pmod{d}$$

will uniquely determine an appropriate $B$ modulo $2a_1a_2$.

We are grateful to Jerome Solinas for detecting and to Robert L. Ward for correcting an error in the proof of this proposition in an earlier version.

Now, if we have united forms $(a_1,\ B,\ a_2 C)$ and $(a_2,\ B,\ a_1 C)$, then under the transformation

$$\begin{pmatrix} X \\ Y \end{pmatrix} = \begin{pmatrix} 1 & 0 & 0 & -C \\ 0 & a_1 & a_2 & B \end{pmatrix} \begin{pmatrix} x_1 x_2 \\ x_1 y_2 \\ y_1 x_2 \\ y_1 y_2 \end{pmatrix} \tag{4.2}$$

we have the equation

$$(a_1 x_1^2 + B x_1 y_1 + a_2 C y_1^2)(a_2 x_2^2 + B x_2 y_2 + a_1 C y_2^2) = \tag{4.3}$$
$$a_1 a_2 X^2 + BXY + CY^2.$$

We define the *form compounded of* $f_1$ *and* $f_2$ to be $(a_1 a_2,\ B,\ C)$, which we write $f_1 \circ f_2$. [1] A lemma is needed to carry out the next major proof.

---

[1] As with the word *anceps*, the translation from the Latin of Gauss has been imperfect. There is total agreement on the use of *composition* as the noun, but both *to compound* and *to compose* can be found as the infinitive.

**Lemma 4.6.** *Two forms* $(a_1,\ b_1,\ c_1)$ *and* $(a_2,\ b_2,\ c_2)$ *of the same discriminant are equivalent if and only if integers* $\alpha$ *and* $\gamma$ *can be found such that*

$$a_1\alpha^2 + b_1\alpha\gamma + c_1\gamma^2 = a_2$$
$$2a_1\alpha + (b_1 + b_2)\gamma \equiv 0 \pmod{2a_2}$$
$$(b_1 - b_2)\alpha + 2c_1\gamma \equiv 0 \pmod{2a_2}.$$

**Proof.** If the forms are equivalent, then clearly $\alpha$ and $\gamma$ exist with which to represent $a_2$, by equation (1.2). There are infinitely many choices for $\beta$ and $\delta$. For all of them, however, we have the following:

$$\alpha\delta - \beta\gamma = 1$$
$$(b_1\gamma + 2a_1\alpha)\beta + (b_1\alpha + 2c_1\gamma)\delta = b_2.$$

We have two linear equations in the two unknowns $\beta$ and $\delta$. The congruences desired follow from solving for $\beta$ and $\delta$:

$$2a_1\alpha + (b_1 + b_2)\gamma = 2a_2\delta$$
$$(b_1 - b_2)\alpha + 2c_1\gamma = -2a_2\beta.$$

The proof in the opposite direction is simply a reversal of the process of solving the linear equations.

We have defined composition and now must show that it possesses the desired properties. This proof follows that of Dirichlet.

**Theorem 4.7.** *If* $f_1$ *and* $f_2$ *are united forms, and united forms* $f_3$ *and* $f_4$ *exist for which* $f_1 \sim f_3$ *and* $f_2 \sim f_4$, *then* $f_1 \circ f_2 \sim f_3 \circ f_4$.

**Proof.** We may assume that we have forms

$$f_1 = (a_1,\ B,\ a_2C)$$
$$f_2 = (a_2,\ B,\ a_1C)$$
$$f_3 = (m_1,\ N,\ m_2L)$$
$$f_4 = (m_2,\ N,\ m_1L).$$

We define integers $x_1$, $x_2$, $y_1$, and $y_2$ by Lemma 4.6 so that

$$a_1x_1^2 + Bx_1y_1 + a_2Cy_1^2 = m_1$$
$$2a_1x_1 + (B+N)y_1 \equiv 0 \pmod{2m_1} \qquad (4.4)$$
$$(B-N)x_1 + 2a_2Cy_1 \equiv 0 \pmod{2m_1}$$

$$a_2x_2^2 + Bx_2y_2 + a_1Cy_2^2 = m_2$$
$$2a_2x_2 + (B+N)y_2 \equiv 0 \pmod{2m_2} \qquad (4.5)$$
$$(B-N)x_2 + 2a_1Cy_2 \equiv 0 \pmod{2m_2}.$$

We need to show that integers $X$ and $Y$ exist such that

$$a_1a_2X^2 + BXY + CY^2 = m_1m_2$$
$$2a_1a_2X + (B+N)Y \equiv 0 \pmod{2m_1m_2} \qquad (4.6)$$
$$(B-N)X + 2CY \equiv 0 \pmod{2m_1m_2}.$$

Clearly, the equation is satisfied. By direct calculation (the crucial step is the substitution $N^2 = B^2 - 4a_1a_2C + 4m_1m_2L$) one can show that

$$(a_1x_1 + (B+N)y_1/2)(a_2x_2 + (B+N)y_2/2) \equiv$$
$$a_1a_2X + (B+N)Y/2 \pmod{m_1m_2}.$$

To prove the second congruence, we note that if we write

$$(B - \sqrt{\Delta})X/2 + CY = U,$$

then the following four equations are immediate:

$$
\begin{aligned}
\left[(B - \sqrt{\Delta})x_1/2 + a_2Cy_1\right]\left[a_2x_2 + (B + \sqrt{\Delta})y_2\right] &= a_2U \\
\left[a_1x_1 + (B + \sqrt{\Delta})y_1\right]\left[(B - \sqrt{\Delta})x_2/2 + a_1Cy_2\right] &= a_1U \\
\left[(B - \sqrt{\Delta})x_1/2 + a_2Cy_1\right]\left[(B - \sqrt{\Delta})x_2/2 + a_1Cy_2\right] &= (B - \sqrt{\Delta})U/2 \\
C\left[a_1x_1 + (B + \sqrt{\Delta})y_1\right]\left[a_2x_2 + (B + \sqrt{\Delta})y_2\right] &= (B + \sqrt{\Delta})U/2
\end{aligned}
$$

As above, we can substitute $N$ for $\sqrt{\Delta}$ and convert these four equa-tions into congruences modulo $m_1m_2$. Since the left-hand sides are all congruent to 0 modulo $m_1m_2$, and the forms are assumed to be united, we must have $U \equiv 0 \pmod{m_1m_2}$, which is the desired second con-gruence.

The essence of Theorem 4.7 is that the class determined by the composition of two individual forms depends not on the forms but only on their classes. Composition is easily seen to be commutative and associative. Further, for any forms $(1, b_1, c_1)$ and $(a_2, b_2, c_2)$ we have

$$(1, b_1, c_1) \circ (a_2, b_2, c_2) \sim (a_2, b_2, c_2).$$

Finally, we note that for any form $(a, b, c)$ we have

$$(a, b, c) \circ (a, -b, c) \sim (a, b, c) \circ (c, b, a) \sim (ac, b, 1).$$

We have proved the following theorem, of enormous importance in the theory of numbers.

**Theorem 4.8.** *Under composition, the classes of forms of a fixed discriminant form a finite abelian group. The identity of the group is the principal class, and the inverse of the class of any form is the class of the opposite of the form.*

There is a further important fact which shall be stated formally.

**Corollary 4.9.** *The classes which are of order 2 in the class group are precisely those classes which contain ambiguous forms.*

**Proof.** An ambiguous form is equivalent to its own opposite.

We have defined the substitutions (4.2) for forms after determining the correct center coefficient $B$. It is an easy matter to show that the following is the analogous substitution if we begin with united forms $(a_1, b_1, c_1)$ and $(a_2, b_2, c_2)$ in which the center coefficient has not yet been adjusted.

$$\begin{pmatrix} X \\ Y \end{pmatrix} = \begin{pmatrix} 1 & \frac{b_2-B}{2a_2} & \frac{b_1-B}{2a_1} & \frac{[b_1b_2+\Delta-B(b_1+b_2)]}{4a_1a_2} \\ 0 & a_1 & a_2 & \frac{b_1+b_2}{2} \end{pmatrix} \begin{pmatrix} x_1x_2 \\ x_1y_2 \\ y_1x_2 \\ y_1y_2 \end{pmatrix}$$

With this substitution with integer coefficients, we have the following analog of (4.3):

$$(a_1x_1^2 + b_1x_1y_1 + c_1y_1^2) \cdot (a_2x_2^2 + b_2x_2y_2 + c_2y_2^2) =$$
$$a_1a_2X^2 + BXY + \left( \frac{B^2 - \Delta}{4a_1a_2} \right) Y^2.$$

Although the concept of united forms is useful for an initial description of the class group, it is less useful for computational purposes, not being algorithmic. We present the composition algorithm of Arndt, as described in Mathews. (Since Mathews uses a definition of forms different from ours, the algorithm has been modified accordingly.)

**Theorem 4.10.** *To compound* $f_1 = (a_1,\ b_1,\ c_1)$ *and* $f_2 = (a_2,\ b_2,\ c_2)$ *let* $\beta = (b_1 + b_2)/2$. *Let* $n = \gcd(a_1, a_2, \beta)$, *and choose* $t$, $u$, $v$ *such that* $a_1 t + a_2 u + \beta v = n$. *Then let* $A = a_1 a_2 / n^2$ *and*

$$B = \frac{a_1 b_2 t + a_2 b_1 u + v(b_1 b_2 + \Delta)/2)}{n}.$$

*Then* $B$ *is a solution of the simultaneous congruences*

$$\frac{a_1 B}{n} \equiv \frac{a_1 b_2}{n} \pmod{2A}$$

$$\frac{a_2 B}{n} \equiv \frac{a_2 b_1}{n} \pmod{2A}$$

$$\frac{\beta B}{n} \equiv \frac{b_1 b_2 + \Delta}{2n} \pmod{2A}.$$

*The form compounded of* $f_1$ *and* $f_2$ *is then*

$$(\frac{a_1 a_2}{n^2},\ B,\ *),$$

*the third coefficient being determined from the discriminant formula.*

**Proof.** Let $n$, $t$, $u$, and $v$ be computed as above. Since we have $\Delta = b_1^2 - 4a_1 c_1 = b_2^2 - 4a_2 c_2$, we now have

$$\left(\frac{b_1 + b_2}{2}\right) \cdot \left(\frac{b_1 - b_2}{2}\right) = a_2 c_2 - a_1 c_1$$

and then

$$\left(\frac{b_1 + b_2}{2n}\right) \cdot \left(\frac{b_1 - b_2}{2}\right) = \frac{a_2 c_2}{n} - \frac{a_1 c_1}{n}.$$

Any further common factors of $a_1$ and $a_2$ must divide $(b_1 - b_2)/2$ and be prime to $(b_1 + b_2)/2$. It is therefore possible to find $B$ such that

$$B \equiv b_1 \pmod{2a_1/n}$$
$$\equiv b_2 \pmod{2a_2/n}.$$

By carefully arguing further as in the proof of Proposition 4.5, we obtain a substitution

$$\begin{pmatrix} X \\ Y \end{pmatrix} = \begin{pmatrix} n & \frac{(b_2-B)n}{2a_2} & \frac{(b_1-B)n}{2a_1} & \frac{[b_1 b_2 + \Delta - B(b_1+b_2)]n}{4a_1 a_2} \\ 0 & \frac{a_1}{n} & \frac{a_2}{n} & \frac{b_1+b_2}{2n} \end{pmatrix} \begin{pmatrix} x_1 x_2 \\ x_1 y_2 \\ y_1 x_2 \\ y_1 y_2 \end{pmatrix}$$

with integer coefficients. The proof now follows exactly as the original proof for composition of united forms.

**Proposition 4.11.** *As defined above,*

$$B = b_1 + \left( \frac{2a_1}{n} \right) \cdot \left( \frac{t(b_2 - b_1)}{2} - c_1 v \right).$$

**Proof.**

$$
\begin{aligned}
n(B - b_1) &= -nb_1 + a_1 b_2 t + a_2 b_1 u + v(b_1 b_2 + \Delta)/2 \\
&= b_1(a_2 u - n) + a_1 b_2 t + v(b_1 b_2 + \Delta)/2 \\
&= b_1(-a_1 t - \beta v) + a_1 b_2 t + v(b_1 b_2 + \Delta)/2 \\
&= a_1 t(b_2 - b_1) + v((b_1 b_2 + \Delta)/2 - b_1 \beta) \\
&= a_1 t(b_2 - b_1) + v(b_1 b_2 + b_1^2 - 4a_1 c_1 - b_1^2 - b_1 b_2)/2 \\
&= a_1 t(b_2 - b_1) + v(-4a_1 c_1)/2 \\
&= a_1 t(b_2 - b_1) - 2a_1 c_1 v \\
&= 2a_1(t(b_2 - b_1)/2 - c_1 v).
\end{aligned}
$$

The proposition is proved.

We present a final composition algorithm, which is not unlike Arndt's method. This algorithm for use in machine computation was first stated by Daniel Shanks [SHAN69].

**Theorem 4.12.** *To compound* $f_1 = (a_1, b_1, c_1)$ *and* $f_2 = (a_2, b_2, c_2)$, *let* $\beta = (b_1 + b_2)/2$. *Let* $m = \gcd(a_1, \beta)$, *and* $n = \gcd(m, a_2)$. *Solve* $a_1 x + \beta y = m$ *for* $x$ *and* $y$ *and*

$$mz/n \equiv x\left(\frac{b_2 - b_1}{2}\right) - c_1 y \pmod{a_2/n} \ \text{for } z.$$

*The form compounded of* $f_1$ *and* $f_2$ *is then*

$$(a_1 a_2/n^2, \ b_1 + 2a_1 z/n, \ *),$$

*with the third coefficient being computed from the discriminant formula.*

**Proof.** We let $a_1 x + \beta y = m$, and $mp + a_2 u = n$. Then

$$a_1 px + \beta py + a_2 u = n,$$

so that $t = px$, $u = u$, and $v = py$ suffices for the choice of $t$, $u$, $v$ in Arndt's method. By Proposition 4.11, the chosen value $B$ in Arndt's method is

$$B = b_1 + \left(\frac{2a_1}{n}\right) \cdot \left[px\left(\frac{b_2 - b_1}{2}\right) - c_1 py\right].$$

We note that $pm/n$, which is an integer, is congruent to 1 $\pmod{a_2/n}$ so that

$$z \equiv pmz/n \equiv px\left(\frac{b_2 - b_1}{2}\right) - c_1 py \pmod{a_2/n}.$$

Different choices of $z$ thus produce values of $B$ differing by a multiple of $2a_1a_2/n^2$, that is, forms equivalent to those chosen by Arndt's method.

An important special case of composition is *duplication* of a form, that is to say, squaring the class.

**Corollary 4.13.** *To compound $f = (a, \ b, \ c)$ with itself, let $n = \gcd(a, b)$, and solve $by/n \equiv 1 \pmod{a/n}$ for $y$. Then $(a, \ b, \ c) \circ (a, \ b, \ c) \sim (a^2/n^2, \ b - 2acy/n, \ *)$, with the third coefficient computed from the discriminant formula.*

**Proof.** We apply the algorithm of Theorem 4.12, noting that many of the terms drop out.

## 4.3   Generic Characters Revisited

Let $G$ be a finite abelian group.  A function $\chi : G \to \mathbf{C}$ is a *group character* if

a) for some $A \in G$, $\chi(A) \neq 0$;

b) for all $A, B \in G$, $\chi(AB) = \chi(A) \cdot \chi(B)$.

We call a character $\chi_1$ the *principal character* if $\chi_1(A) = 1$ for all $A \in G$.

**Proposition 4.14.** *Let $G$ be a finite abelian group, and let $\chi$ be any character.*

*a) If $1_G$ is the identity of $G$, then $\chi(1_G) = 1$.*

*b) If $|G| = h$, then $\chi(A)$ is some $h$-th root of unity.*

*c)*

$$\sum_{A \in G} \chi(A) \ = \ |G|, \ \textit{if } \chi = \chi_1,$$
$$= \ 0, \ \textit{otherwise.}$$

**Proof.** a) For all $A$, $\chi(A) = \chi(A \cdot 1_G) = \chi(A) \cdot \chi(1_G)$. The character cannot be identically zero, by definition, so there is some $A$ for which $\chi(A)$ is not zero.

b) If $h$ is the order of the group $G$, then for any element $A$, we have $A^h = 1_G$. Thus $1 = \chi(1_G) = \chi(A^h) = \chi(A)^h$.

c) Finally, if $\chi$ is the principal character, part c is obvious. If not, let $S$ be the sum in question. Then for any $B \in G$ for which $\chi(B) \neq 0$, we have

$$S \cdot \chi(B) = \sum_{A \in G} \chi(A)\chi(B) = \sum_{A \in G} \chi(AB) = S.$$

Since, then, we have $S \cdot (\chi(B) - 1) = 0$, we must have $S = 0$.

The next proposition is obvious from the definition of group character and a direct computation showing the multiplicativity of the generic characters.

**Proposition 4.15.** *The generic characters of any class group are group characters.*

**Theorem 4.16.** *For any discriminant $\Delta$, each genus has the same number of classes of forms. Composition of classes of forms is a well-defined operation on the genera, which form a group under composition. The number of genera is a power of 2.*

**Proof.** Each of the generic characters is a group character. If the group formed, for each class group, is viewed as the direct product of its generic characters, then the mapping of an element of the class group into the $n$-tuple of $+1$'s and $-1$'s which are its generic characters, is a homomorphism. The genera are the cosets under the homomorphism and thus have the same number of elements (classes) as the principal genus, by Lagrange's theorem. The second statement of the theorem is merely the group isomorphism theorem. The quotient formed from any group modulo the kernel of a homomorphism is a group isomorphic to the group which is the image of the homomorphism. The last statement follows from considering that the group of genera is abelian and consists

only of elements whose square is the identity. It is thus a direct product
of groups of two elements, and has an order which is some power of 2.

**Proposition 4.17.** *The number of ambiguous classes (including the
principal class) is equal to one-half the number of possible genera.*

**Proof.** *Case i.* $\Delta$ is odd and negative. If $\Delta$ has $k$ prime factors,
then there are $2^{k-1}$ factorings of $\Delta = -ac$ with $a$ and $c$ coprime and
$0 < a < c$. We thus have $2^{k-1}$ ambiguous forms $(a, a, (a+c)/4)$,
noting that

$$(a,\ a,\ (a+c)/4) \sim ((a+c)/4,\ -a,\ a) \sim ((a+c)/4,\ c,\ c) \sim$$
$$(c,\ -c,\ (a+c)/4) \sim (c,\ c,\ (a+c)/4).$$

*Case ii.* $\Delta$ is odd and positive. We get twice as many ambiguous

forms since we can now choose both positive and negative coefficients.
But by a previous theorem, each ambiguous cycle contains exactly two
ambiguous forms, so we are again left with $2^{k-1}$ ambiguous classes.
*Case iii.* $\Delta$ is even and negative. We have, for the odd prime factors of
$\Delta$, the same forms as above. In addition, we have the following, writing
$\Delta' = \Delta/4$:

   $(2,\ 2,\ c)$ for odd $c$ when $\Delta' = 2c - 1$ is odd;

   $(2,\ 0,\ \Delta'/2)$ for $\Delta' \equiv 2, 6 \pmod 8$;

   $(4,\ 0,\ \Delta'/4)$ for $\Delta' \equiv 4 \pmod 8$;

   $(4,\ 4,\ c)$ for odd $c$ when $\Delta' = 4(c-1) \equiv 0 \pmod 8$;

   $(8,\ 0,\ \Delta'/8)$ for $\Delta' \equiv 0 \pmod 8$.

A careful examination of the small discriminants and an exhaustive
check of these cases show that these provide, with the products with
the other ambiguous forms, exactly as many as are needed.

*Case iv.* $\Delta$ is even and positive. We shall not do this case, as it is similar to Case iii, but requires yet more checking of cases.

We now restrict ourselves for the moment to fundamental discriminants. A discriminant $\Delta$ of quadratic forms is called *fundamental* if either one of the following holds:

1. $\Delta$ is odd and squarefree;

2. $\Delta$ is even, $\Delta/4$ is squarefree, and $\Delta/4 \equiv 2$ or $3 \pmod 4$.

We note first that odd fundamental discriminants have only the generic characters $\left(\frac{m}{p}\right)$ for the primes $p$ which divide the discriminant. Even fundamental discriminants $\Delta$ have the additional character $\chi$, $\psi$, or $\chi \cdot \psi$, according as $\Delta$ is congruent to 3 modulo 4 or to 2 or 6 modulo 8.

**Proposition 4.18.** *Let $\Delta$ be a fundamental discriminant. Then the product of the assigned values for the characters for any given genus is $+1$.*

**Proof.** We first do the case of odd discriminants $\Delta$. Since $\Delta$ is necessarily congruent to 1 modulo 4, there are an even number $2k$ of prime factors of $\Delta$ congruent to 3 modulo 4. Then, if $(a, b, c)$ is a form of discriminant $\Delta$, with $a$ odd and relatively prime to $\Delta$, we have

$$+1 = \left(\frac{\Delta}{a}\right) = \prod \left(\frac{p}{a}\right) = \prod_{p \equiv 1 \ (\text{mod } 4)} \left(\frac{p}{a}\right) \cdot \prod_{p \equiv 3 \ (\text{mod } 4)} \left(\frac{p}{a}\right) =$$

$$\chi(a)^{2k} \cdot \prod_{p \equiv 1 \ (\text{mod } 4)} \left(\frac{a}{p}\right) \cdot \prod_{p \equiv 3 \ (\text{mod } 4)} \left(\frac{a}{p}\right) = \prod \left(\frac{a}{p}\right),$$

where the products run over the primes $p$ which divide $\Delta$. This final product is the product of the generic characters, and the proposition is proved in this case.

Next we consider even discriminants $\Delta$, for which $\Delta/4$ is odd. Since $\Delta/4$ is necessarily congruent to 3 modulo 4, there are an odd number $2k + 1$ of prime factors of $\Delta$ congruent to 3 modulo 4. The proof is now a repetition of the argument above, except that the needed factor of $\chi(a)$ remains.

Finally, we consider discriminants congruent to 0 modulo 8. As above, we have, if $(a, \ b, \ c)$ is a form of discriminant $\Delta$, with $a$ odd and relatively prime to $\Delta$,

$$+1 = \left(\frac{\Delta}{a}\right) = \left(\frac{\Delta/4}{a}\right) = \left(\frac{2}{a}\right) \cdot \prod \left(\frac{p}{a}\right) =$$

$$\psi(a) \cdot \prod_{p \equiv 1 \pmod 4} \left(\frac{p}{a}\right) \cdot \prod_{p \equiv 3 \pmod 4} \left(\frac{p}{a}\right) =$$

$$\psi(a) \cdot \chi(a)^k \cdot \prod_{p \equiv 1 \pmod 4} \left(\frac{a}{p}\right) \cdot \prod_{p \equiv 3 \pmod 4} \left(\frac{a}{p}\right) =$$

$$\psi(a) \cdot \chi(a)^k \cdot \prod \left(\frac{a}{p}\right),$$

where the products run over the odd primes $p$ which divide $\Delta$ and $k$ is as before. Now, if $\Delta/4 \equiv 2 \pmod 8$, and $\psi$ is a character, then $k$ is even and $\chi(a)^k$ is $+1$. If $\Delta/4 \equiv 6 \pmod 8$, and $\chi \cdot \psi$ is a character, then $k$ is odd and $\chi(a)^k = \chi(a)$. In either case, the proposition is seen to be true.

The restriction that $\Delta$ be fundamental is necessary in Proposition 4.18, as the example of discriminant $-1520$ shows.

**Theorem 4.19.** *If $\Delta$ is a fundamental discriminant, then exactly half of the possible genera exist.*

**Proof.** By the previous proposition, we see that at most half of the genera could exist, namely those with an even number of $-1$ characters.

To prove that these must appear, we appeal to Dirichlet's theorem that in any "reasonable" congruence class, primes appear infinitely often (though we need only one).

**Theorem.** *(Dirichlet) If* $\gcd(a, b) = 1$, *then an infinite number of primes appear in the linear sequence* $\{a + bx : x \in \mathbf{Z}\}$.

Given a set of generic characters possible in light of Proposition 4.18, we have a list of possible congruence classes into which $a$ must fall modulo 4, 8, and/or odd prime factors of $\Delta$. Using the Chinese Remainder Theorem, we can put this into a congruence of the form of Dirichlet's theorem, and find a prime $a$ which satisfies these congruences. Since $a$ is a prime, we can actually guarantee the solvability of $b^2 \equiv \Delta \pmod{4a}$ and get a form of the desired genus.

**Theorem 4.20.** *Let* $\Delta$ *be a fundamental discriminant. Then*

a) *if* $\Delta$ *has one generic character, then its assigned value is* $+1$;

b) *if* $\Delta$ *has two characters, then the assigned values are* $(+1, +1)$ *and* $(-1, -1)$;

c) *if* $\Delta$ *has more than two characters, then the product of the assigned values of any given character is* $+1$.

**Proof.** Parts a and b are obvious, given the above theorems. To prove part c, we consider any $\Delta$ with at least three characters, that is, at least four genera, and prove a result somewhat stronger. We shall prove that in any subgroup of genera of order four or larger, the product of the assigned values of any given character $\lambda$ is $+1$. We start with the principal genus $G_1$; clearly, $\lambda(G_1) = +1$. We then choose any other two genera $G_2$ and $G_3$. The set $\{G_1, G_2, G_3, G_2G_3\}$ forms a Klein four-group, and either none or two of the values $\lambda(G_1)$, $\lambda(G_2)$, $\lambda(G_3)$, $\lambda(G_2G_3)$ are $-1$. If there are no more genera, the theorem is proved. If there is another genus $G_4$, then if $\lambda(G_4)$ is $+1$, then the values $\lambda(G_4)$, $\lambda(G_2G_4)$, $\lambda(G_3G_4)$, and $\lambda(G_2G_3G_4)$ are the same as the values $\lambda(G_1)$, $\lambda(G_2)$, $\lambda(G_3)$, and $\lambda(G_2G_3)$. If not, then the values are reversed. In either case, if this constitutes all the genera, the theorem is proved. If not, we may continue in this fashion until the group of genera is exhausted.

We state one final theorem, whose proof will be based on a major theorem we shall not prove.

**Theorem 4.21.** *The principal genus consists exactly of the subgroup of squares of classes of forms.*

**Proof.** It is clear that the square of any class is in the principal genus. To show that any class in the principal genus is the square of some class, we cite the decomposition theorem for finite abelian groups.

**Theorem.** *If $G$ is any finite abelian group, then $G$ can be written as a direct product of cyclic groups*

$$C(2^{i_1}) \times C(2^{i_2}) \times \cdots C(2^{i_j}) \times C(r_1) \times C(r_2) \times \cdots C(r_k),$$

*where $C(n)$ is a cyclic group of order $n$, the $i_l$ are positive, and the $r_l$ are all odd.*

From the various results above it can be seen that the ambiguous forms for any given discriminant generate a group which is the direct product of $j$ groups of order 2 and that there are $2^j$ genera. Since any element of odd order is a square, it is then evident that the number of squares in such a group is

$$2^{i_1-1} \cdot 2^{i_2-1} \cdots 2^{i_j-1} \cdot r_1 \cdot r_2 \cdots r_k.$$

This happens to be the number of classes in any given genus, since the mapping of classes to their characters is a homomorphism. Thus all the classes in the principal genus must necessarily be squares.

## 4.4   Representation of Integers

We can now return to the questions posed at the beginning of Chapter 1, namely:

a) What integers can be represented by a given form?

b) What forms can represent a given integer?

c) If a form represents an integer, how many representations exist, and how may they all be found?

The answer for forms of discriminant $-4$ was given in Chapter 1 and can be proved without resorting to the theory of forms. The answer is, however, a model for the theorems which shall now be stated. We shall not present proofs of these since they are merely the collection of the results on representation obtained up to this point. Although we have thus far dealt only with primitive forms, this will be rephrased in the statement of these theorems so as to make them self-contained.

**Theorem 4.22.** *If any form $f(x,y) = ax^2 + bxy + y^2$ of discriminant $\Delta$ represents an integer $m$ which is divisible by an odd prime $p$ for which $\left(\frac{\Delta}{p}\right) = -1$, then we must have, for some positive integer $k$, $p^2k \parallel m$, $p^k \parallel x$, and $p^k \parallel y$. No integer which is exactly divisible by an odd power of a prime $p$ for which $\left(\frac{\Delta}{p}\right) = -1$ can be primitively representable by any primitive form of discriminant $\Delta$.*

**Theorem 4.23.** *If $p$ is an odd prime for which $\left(\frac{\Delta}{p}\right) = +1$, and $b$ is any solution to $b^2 \equiv \Delta \pmod{4p}$, then $p$ is (primitively) representable by the forms belonging to the classes of $(p, b, *)$ and $(p, -b, *)$, which are distinct, and by no other forms.*

**Theorem 4.24.** *If $p$ is an odd prime which divides $\Delta$, then $p$ is (primitively) representable by the forms belonging to the class of $(p, 0, *)$ or $(p, p, *)$, depending on the parity of $\Delta$, and by no other forms.*

**Theorem 4.25.** *The prime 2 is representable by the forms belonging to the classes of $(2, 1, *)$ and $(2, -1, *)$, for odd discriminants $\Delta \equiv 1$ (mod 8), and by no other forms, and is not representable by any form of any discriminant $\Delta \equiv 5$ (mod 8). The prime 2 is representable by all forms in the single class of discriminant $-4$. For all other even discriminants, 2 is representable by the forms belonging to the classes of $(2, 0, *)$ or $(2, 2, *)$, according as $\Delta \equiv 0$ (mod 8) or $\Delta \equiv 4$ (mod 8), and by no other forms.*

**Theorem 4.26.** *If $m$ is any integer and $\Delta$ any discriminant, then $m$ can be primitively represented by the forms belonging to the classes $(m, \pm b, *)$, where $b$ is any solution of the congruence $b^2 \equiv \Delta$ (mod $4m$), and by no other forms. If we have $m = \prod p_i^{\alpha_i}$ as the canonical factoring of $m$ into prime powers, then the set of these classes is exactly the set of classes obtained by representing the primes $p_i$ as in Theorems 4.24-4.26 with forms $f_i$ and computing the composition $\prod f_i^{\alpha_i}$ in the class group of discriminant $\Delta$.*

**Theorem 4.27.** *All the representations of integers in Theorems 4.22-4.26 can be obtained by applying the transformations of the modular group to the obvious representations with $(x, y) = (\pm 1, 0)$.*

# Chapter 5

# Miscellaneous Facts

## 5.1 Class Number Computations

Many of the groups listed in Chapters 2 and 3 can be described completely without recourse to the composition algorithm. For class numbers 1, 2, and 3, only cyclic groups are possible. For class number 4, only the cyclic or Klein 4-groups are possible, and these can be distinguished by the number of ambiguous forms. Thus, the groups for discriminants $-39$, $-55$, $-63$, $-155$, $-56$, $-68$, and $-80$ are cyclic, while the groups of discriminants $-84$ and $-96$ are not. The structure of these groups then determines the multiplication table completely.

The first occurrence of an "interesting" group for negative discriminant is for discriminant $-47$, with class number 5. It is easy to see from the united forms definition that

$$(2,\ 1,\ 6) \circ (2,\ 1,\ 6) \sim (4,\ 1,\ 3) \sim (3,\ -1,\ 4).$$

Thus, the group is a 5-cycle representable as

$$\begin{aligned} identity &= (1,\ 1,\ 12) \\ a &= (2,\ 1,\ 6) \end{aligned}$$

$$a^2 = (3, -1, 4)$$
$$a^3 = (3, 1, 4)$$
$$a^4 = (2, -1, 6).$$

It is easy to produce a class group with a large number of ambiguous forms and thus a group that is nontrivial in the sense that it has a large number of classes of order 2. It is only necessary to consider negative discriminants with a large number of distinct prime factors, such as $-5460 = -4 \cdot 3 \cdot 5 \cdot 7 \cdot 13$, with reduced ambiguous forms

$$(1, 0, 1365)$$
$$(3, 0, 455)$$
$$(5, 0, 273)$$
$$(7, 0, 195)$$
$$(13, 0, 105)$$
$$(15, 0, 91)$$
$$(21, 0, 65)$$
$$(35, 0, 39).$$

The multiplicative structure of this subgroup of order 8 should be completely obvious. It is, however, more difficult, as we shall see, to produce groups with noncyclic subgroups of orders 9, 25, 49, and so forth. The first such examples of even and odd discriminants are $-3299$, $-3896$, $-11199$, $-17944$, $-63499$, and $-159592$, of class numbers 27, 36, 100, 50, 49, and 98, respectively.

For positive discriminants, the groups are much simpler. The first discriminants with class numbers 3, 4, 5, 6, 7, and 8 are 229, 105, 401, 321, 577, 505, 60, 316, and 780.

Aside from the exceptional discriminants of very small magnitude, an odd class number can only occur for an odd prime discriminant. Also, as a general rule, the class groups for negative discriminants tend to grow in order and complexity, not really as a function of the discriminant but of the *radicand*, which is the discriminant $\Delta$ for odd discriminants and $\Delta/4$ for even discriminants. That is, the class groups for even discriminants of magnitude approximately $M$ will resemble the class groups for odd discriminants, not of discriminants of magnitude $M$, but of magnitude $M/4$.

As stated earlier, a discriminant $\Delta$ of quadratic forms is called *fundamental* if either one of the following holds:

1. $\Delta$ is odd and squarefree;

2. $\Delta$ is even, $\Delta/4$ is squarefree, and $\Delta \equiv 2$ or 3 (mod 4).

The class numbers for negative fundamental discriminants have been computed to discriminant $-25000000$ [BUEL76, BUEL84, BUEL87a]. The class numbers for positive fundamental discriminants have been computed to a lesser limit, but at least to 4000000. For negative discriminants the computation is not difficult. To compute the class number for a single small discriminant, the direct approach is no doubt best. The bound $0 \leq |b| \leq \sqrt{D/3}$ provides a list of potential values for $b$ and thus a list of potential values $ac = (b^2 - \Delta)/4$. From here the conditions $|b| \leq a \leq c$ and $\gcd(a, b, c) = 1$ limit the number of choices of classes of forms, and they can be produced explicitly. For the computation of class numbers of all discriminants in a given range, an indirect method [LEHM69] is preferable. First of all, the class numbers for even discriminants should be computed separately from those for odd discriminants. The technique is then to set aside an array in memory, with the subscript in the array representing the discriminant. For a given range of discriminants, a maximum value of the center coefficient $b$ can be computed, and from these lower and upper limits on possible values of $a$ and $c$ for reduced forms can be determined. A triple loop, on positive values of $b$, $a$, and $c$, then counts all the reduced forms whose discriminants fall within the range of discriminants considered. The only complication is to take care at the endpoints of the loops, counting the ambiguous forms only once rather than twice.

Computation of class numbers for positive discriminants is easily recognized to be much more complicated. In the first place, storage needs are greater. For negative discriminants, the reduced forms need never be saved, only counted. For positive discriminants, the forms must be saved so they can later be grouped into cycles. The numbers of reduced forms of positive and of negative discriminant of similar magnitude are comparable, but the class numbers differ greatly. As a general rule, the computations in groups of negative discriminant are dominated by the necessary complexity of a group, even a cyclic group, with a large number of elements. The computations in groups of positive discriminants are dominated by the required cycling necessary for recognizing that two reduced forms are equivalent. In the worst case we can either be forced to pass through the entire cycle to find a match or to implement some more sophisticated technique of storing, say, forms of small coefficients and comparing the lists by some sort of matching algorithm.

# 5.2 Extreme Cases and Asymptotic Results

Among the celebrated problems of number theory are these:

Which discriminants have class number 1?

Which discriminants have one class per genus?

In the case of negative discriminants, the answers are known. There are, in fact, nine negative fundamental discriminants of class number one: $-3$, $-4$, $-7$, $-8$, $-11$, $-19$, $-43$, $-67$, $-163$. Gauss, who knew all of these, conjectured that no more existed. In the 1930's Heilbronn and Linfoot [HEIL34] proved that at most one more existed. A proof by Heegner [HEEG] was thought to contain a gap [STAR69], and Harold Stark [STAR67] and Alan Baker [BAKE66] independently proved (again) in the late 1960's that there were no more. It is thus also known (for reasons which will appear in Chapter 7, when fundamental and nonfundamental discriminants are compared) that there are four nonfundamental discriminants of class number one: $-12$, $-16$, $-27$, $-28$. Finally, we present two tables, of the remaining 56 fundamental discriminants of one class per genus and of the remaining 32 nonfundamental discriminants of one class per genus:

| | | | | | |
|---|---|---|---|---|---|
| $-15$ | $-115$ | $-235$ | $-427$ | $-708$ | $-1380$ |
| $-20$ | $-120$ | $-267$ | $-435$ | $-715$ | $-1428$ |
| $-24$ | $-123$ | $-280$ | $-483$ | $-760$ | $-1435$ |
| $-35$ | $-132$ | $-312$ | $-520$ | $-795$ | $-1540$ |
| $-40$ | $-148$ | $-340$ | $-532$ | $-840$ | $-1848$ |
| $-51$ | $-168$ | $-372$ | $-555$ | $-1012$ | $-1995$ |
| $-52$ | $-187$ | $-403$ | $-595$ | $-1092$ | $-3003$ |
| $-84$ | $-195$ | $-408$ | $-627$ | $-1155$ | $-3315$ |
| $-88$ | $-228$ | $-420$ | $-660$ | $-1320$ | $-5460$ |
| $-91$ | $-232$ | | | | |

| | | | | | |
|---|---|---|---|---|---|
| −32 | −75 | −160 | −315 | −928 | −2080 |
| −36 | −96 | −180 | −352 | −960 | −3040 |
| −48 | −99 | −192 | −448 | −1120 | −3360 |
| −60 | −100 | −240 | −480 | −1248 | −5280 |
| −64 | −112 | −288 | −672 | −1632 | −7392 |
| −72 | −147 | | | | |

These negative discriminants of one class per genus, which were called *idoneal* (from the Latin for *suitable*) numbers by Euler, have the following unique property. We have seen that an integer $r$ is representable by a form of discriminant $\Delta$ if and only if its generic characters for discriminant $\Delta$ match the assigned characters of some genus. We cannot, however, specify which form (technically, which class) represents $r$, and indeed more than one form in the same genus can represent $r$. For the idoneal numbers, however, all representations of $r$ by forms of discriminant $\Delta$ are by the unique class in the given genus. This, coupled with the fact that for positive discriminants the representation problem is elliptic and hence readily bounded (by $\mid y \mid \leq 2\sqrt{ar/\Delta}$), led Gauss to the discovery of a rather elegant algorithm for factoring (which will be discussed in Chapter 10) which would, were there infinitely many idoneal numbers, make the factoring problem much easier.

In the case of positive discriminants, the answers to these two questions are still not known. Computations of class numbers have shown that about 80% of the class numbers for positive prime discriminants are actually 1, and somewhat more than 60% of positive discriminants have one class per genus. It is conjectured, and is almost assuredly true, that the number of discriminants of class number 1 is infinite. This would also imply that the number of discriminants of one class per genus is infinite. This problem, however, is intimately connected with several other deep unsolved problems in number theory, some of which will be mentioned later in this work.

Another aspect of the class number question is worth mentioning at this point, although it will be studied later in more detail. We have mentioned that the numbers of reduced forms of positive and of negative discriminants of similar magnitude are roughly equal. However, for negative discriminants they are all inequivalent, but for positive discriminants they form cycles and are not inequivalent. This observation can be quantified in the following way, forming a partial answer to the question, "How large is the class number relative to the size of the discriminant?" Let $s$ be a real variable, $\Delta$ be a fundamental discriminant of binary quadratic forms, and let $\chi_\Delta(n)$ be the Jacobi symbol $\left(\frac{\Delta}{n}\right)$ if $n$ and $\Delta$ are relatively prime, zero if they are not. The *Dirichlet L-function* for discriminant $\Delta$ is

$$L(s, \chi_\Delta) = \sum_{n=1}^{\infty} \chi_\Delta(n) n^{-s} = \prod_p \left( \frac{p^s}{p^s - \chi_\Delta(p)} \right),$$

where the infinite product is taken over all primes $p$. Convergence of the series and product is obvious for $s > 1$, and can be shown to hold also for $s > 0$. A famous theorem of Dirichlet is the following.

**Theorem.** *The class number $h$ of a fundamental discriminant $\Delta$ is equal to*

$$\frac{\sqrt{\Delta}}{2 \ln \varepsilon} \cdot L(1, \chi_\Delta)$$

*for positive discriminants $\Delta$, and*

$$\frac{k\sqrt{-\Delta}}{\pi} \cdot L(1, \chi_\Delta)$$

*for negative discriminants $\Delta$.*

In these formulas, $\varepsilon$ is the fundamental solution to equation (3.11), if one exists, or the fundamental solution to (3.2) if not, and $k$ is equal to 2 for $\Delta = -4$, 3 for $\Delta = -3$, and 1 for all other negative values of $\Delta$.

The $L$-functions are connected to the well-known Riemann $\zeta$-function,

$$\zeta(s) = \prod_p \left( \frac{p^s}{p^s - 1} \right),$$

and both functions can be considered to be defined for the complex variable $s$. Among the extensive analytic results concerning the values of the $L$-functions is a theorem of Littlewood [LITT28].

**Theorem.** *Assuming the validity of the Riemann hypothesis for the L-functions, that is, that all zeros of $L(1, \chi_\Delta)$ are to found on the vertical line $s = 1/2$, we have, as $\Delta$ (or $-\Delta$) tends to infinity, the asymptotic bounds*

$$L(1, \chi_\Delta) < \{1 + o(1)\} 2e^\gamma \log\log |\Delta|$$

*and*

$$L(1, \chi_\Delta) > \frac{1}{\{1 + o(1)\}\{12/\pi^2\}e^\gamma \log\log |\Delta|},$$

*where $\gamma$ is Euler's constant.*

These results can be combined to show that the class number for a negative discriminant $\Delta$ is "about" $\sqrt{|\Delta|}/\pi$. Further, the class number for positive discriminants is inversely proportional to the size of the fundamental solution of (3.11) or (3.2). Since the class numbers for negative discriminants have been computed, it has been possible to compute the values of the $L$-functions for those discriminants. This has been done for fundamental discriminants from $-1$ to $-4000000$ [SHAN73a, BUEL77]. The values of $L(1, \chi_\Delta)$ being small, ranging from about 0.2 to about 6.35, the values are heavily influenced by the first term in the product expansion, the term corresponding to the prime 2. For this reason, it is useful to consider three cases: even discriminants, for which the first term is 1; odd discriminants congruent to 5 (mod 8), for which the first term is 2/3; and odd discriminants congruent to 1 (mod 8), for which the first term is 2. We find a corresponding variation in the extreme values of the $L$-functions. For even discriminants the maximum value of the $L$-function, for discriminants between $-4$ and $-4000000$, is approximately 3.3926, for discriminant 3519764 of class number 2026. The minimum is approximately 0.3041, for discriminant 1154008, class number 104. For odd discriminants congruent to 5 (mod 8), the maximum value is approximately 2.2789, for discriminant 3724811, class number 1400; the minimum is approximately 0.1988, for discriminant 990127, class number 63. For odd discriminants congruent to 1 (mod 8), the maximum value is approximately 6.3503, for discriminant 2155919, class number 2968; the minimum is approximately 0.6096, for discriminant 3855223, class number 381.

# Chapter 6

# Quadratic Number Fields

## 6.1 Basic Algebraic Definitions

For the benefit of less experienced readers, we repeat some basic definitions of abstract algebra. A *group* is a pair $(G, +_G)$ in which $G$ is a set and $+_G$ is a closed associative binary operation on $G$ for which the following hold:

a) there exists an element $1_G$ of $G$, called the *identity*, which has the property that for any element $a$ in $G$, we have $a +_G 1_G = 1_G +_G a = a$;

b) for every element $a$ of $G$, there exists an element $b$, called the *inverse* of $a$, for which $a +_G b = b +_G a = 1_G$.

A group is called *commutative*, or *abelian*, if the operation $+_G$ is commutative. The *order* of $(G, +_G)$ is the cardinality of the set $G$. Groups are called *finite* or *infinite* according as this cardinality is finite or infinite.

A *ring* is a triple $(R, +_R, *_R)$ with the following properties:

a) $(R, +_R)$ is an abelian group of order at least 2;

b) $+_R$ is a closed associative binary operation on $R$;

c) for all elements of $R$, $*_R$ distributes over $+_R$.

The operation $+_R$ is usually called "addition "and the operation $*_R$ is usually called "multiplication." The identity under addition is usually called the "zero" of the ring. A *ring with identity* is a ring $R$ for which an identity exists for the operation $*_R$. This identity, if it exists, is usually called the "1" of the ring. A *field* is a ring $R$ for which $(R - \{0\}, *_R)$ is an abelian group. In any ring $R$, a subset $I$ of $R$ is called an *ideal* if, for any $\alpha, \beta \in I$, then for any $\lambda, \mu \in R$, $\lambda\alpha + \mu\beta \in R$.

We state without proof some propositions. To avoid confusion later, we shall begin to call the integers $\mathbf{Z}$ the *rational integers*. This distinction will be necessary, as we shall shortly define *algebraic integers*.

**Proposition 6.1.** *The rational integers $\mathbf{Z}$ form a commutative ring with identity under ordinary addition and multiplication. The additive identity is 0, and the additive inverse of $n$ is $-n$. The multiplicative identity is 1.*

**Proposition 6.2.** *The rational numbers $\mathbf{Q}$ form a field under ordinary addition and multiplication. The multiplicative inverse of $a \neq 0$ is $1/a$.*

**Proposition 6.3.** *For any rational integer $n$, $\mathbf{n} = \{kn : k \in \mathbf{Z}\}$ is an ideal of $\mathbf{Z}$. All ideals of $\mathbf{Z}$ are of this form.*

# 6.2 Algebraic Numbers and Quadratic Fields

An *algebraic number* is a complex number $\alpha$ which satisfies a polynomial equation $f(x) = 0$, in which $f(x)$ is a polynomial in $x$ with rational number coefficients. We can multiply this equation by the least common multiple of the denominators of the coefficients of $f(x)$, and may thus equivalently require that $f(x)$ be a polynomial with rational integer coefficients.

An *algebraic integer* is any algebraic number which satisfies a monic polynomial equation $f(x) = 0$ in which $f(x)$ is a polynomial in $x$ with coefficients that are rational integers.

**Proposition 6.4.** *The only rational numbers that are algebraic integers are the rational integers.*

**Proof.** Let $a/b \in \mathbf{Q}$, with $\gcd(a, b) = 1$. Clearly, since $a/b$ is a solution to $x - a/b = 0$, we know that $a/b$ is an algebraic number, and if a rational integer (that is, if $b = 1$), then $a/b = a$ is an algebraic integer. If we assume that $a/b$ is an algebraic integer and $b > 1$, then we have an equation with rational integer coefficients $c_i$:

$$(a/b)^n + c_{n-1}(a/b)^{n-1} + \ldots + c_1(a/b) + c_0 = 0.$$

Multiplying through by $b^n$ we get

$$a^n + c_{n-1}a^{n-1}b + \ldots + c_1ab^{n-1} + c_0b^n = 0.$$

Since $b$ divides all terms except the first, it must also divide the first. This would, however, imply that $\gcd(a, b) > 1$, which is a contradiction.

We now turn our attention specifically to quadratic number fields. Although many of the terms and theorems are valid or have generalizations to all number fields, we shall prove them only for the quadratic case. This is the case in which there are connections with binary quadratic forms, and in this case the proofs are often much simpler because we can simply and explicitly solve quadratic equations. We define a complex number $\alpha$ to be a *quadratic algebraic number* if it satisfies a polynomial equation

$$ax^2 + bx + c = 0, \tag{6.1}$$

where $a, b$, and $c$ are rational integers.

A *quadratic algebraic integer* is a quadratic algebraic number which satisfies a polynomial equation

$$x^2 + bx + c = 0, \tag{6.2}$$

where $b$ and $c$ are rational integers. The quadratic algebraic numbers are precisely the complex numbers of the form

$$\frac{-b + e\sqrt{d}}{2a}, \tag{6.3}$$

where $a, b, d$, and $e$ are rational integers, and $d$ has no square factors other than 1. (We allow $d$ to be 1 so that the rational numbers are all quadratic algebraic numbers and the rational integers are all quadratic algebraic integers. That $d$ is 1 is equivalent to allowing the polynomial (6.1) to be factorable into two linear polynomial factors.) We shall call $d$ the *radicand* of a quadratic algebraic number $\alpha$. We shall also need the following notation: $\mathbf{Q}(\alpha)$ is the smallest field containing both $\mathbf{Q}$ and $\alpha$.

**Proposition 6.5.** *Let $\alpha$ be any quadratic algebraic number of radicand $d$. Then $\mathbf{Q}(\alpha) = \mathbf{Q}(\sqrt{d}) = \{t + u\sqrt{d} : t, u \in \mathbf{Q}\}$. If $d$ is not a perfect* *square, then* $\mathbf{Q}(\sqrt{d})$ *is a vector space of dimension 2 over the field of scalars* $\mathbf{Q}$ *and has a basis* $< 1, \sqrt{d} >$ *over* $\mathbf{Q}$.

**Proof.** Any element $\zeta$ of $\mathbf{Q}(\alpha)$ is a finite sum of powers of $\alpha$ multiplied by rational numbers. However, since $\alpha$ satisfies a quadratic equation (6.2) with rational integer coefficients, we may substitute $-(b/a)\alpha-c/a$ for $\alpha^2$ and reduce any powers higher than the first by one. Repeating this will eliminate all powers higher than the first. Thus $\mathbf{Q}(\alpha) = \{t + u\sqrt{d} : t, u \in \mathbf{Q}\}$. The same process shows that $\mathbf{Q}(\sqrt{d})$ is the same set. That this set is the vector space as described is equally clear.

Let $d \neq 1$ be a rational integer without square factors (except 1). We define $\Delta$ as follows:

a) if $d \equiv 2$ or $d \equiv 3 \pmod 4$, $\Delta = 4d$;

b) if $d \equiv 1 \pmod 4$, $\Delta = d$.

Now, with $d$ and $\Delta$ as defined, we shall call $\mathbf{Q}(\sqrt{d})$ the *quadratic number field* of radicand $d$ and discriminant $\Delta$. We note that $\mathbf{Q}(\sqrt{d}) = \mathbf{Q}(\sqrt{\Delta})$.

**Proposition 6.6.** *Let* $d \neq 1$ *be a rational integer without square factors (except 1). The set of quadratic algebraic integers of radicand $d$ is*
$$\{a + b\sqrt{d} : a, b \in \mathbf{Z}\}, \text{ if } d \equiv 2, 3 \pmod 4;$$
$$\{(a + b\sqrt{d})/2 : a, b \in \mathbf{Z}, a \equiv b \pmod 2\}, \text{ if } d \equiv 1 \pmod 4.$$
*This set is a ring under ordinary addition and multiplication.*

**Proof.** Considering the above definitions and equations (6.1)-(6.3), we see that a quadratic algebraic number $\alpha$ is a quadratic algebraic integer if and only if $\alpha$ is of the form $(-b + e\sqrt{d})/2$, where $b^2 - 4c = e^2d$. Considering this last equation modulo 4, we have, for $d \equiv 2$ or $d \equiv 3$

ind 4), that $b$ and $e$ must both be even rational integers. If $d \equiv 1$ (mod 4), then $b$ and $e$ must be both even or both odd. And, as in Proposition 6.5, to show that this is a ring really only requires the observation that addition and multiplication are closed on these sets.

We shall call the ring defined in Proposition 6.6 the *ring of integers* of $\mathbf{Q}(\sqrt{d})$, written $\mathcal{O}(\sqrt{d})$, or sometimes just $\mathcal{O}$ if the context is clear. However, the distinct terms "algebraic integer" and "rational integer" shall be retained to avoid any possible confusion. That the "algebraic integers" mentioned are actually "quadratic algebraic integers" should be clear since we will not deal further with any algebraic numbers which are not elements of a quadratic field. For convenience in expression, we shall let $\delta$ be $-\sqrt{d}$ or $(1 - \sqrt{d})/2$ according as the discriminant of the field in question is even or odd. With this notation, the ring of integers $\mathcal{O}(\sqrt{d})$ can then be written as $\{a + b\delta : a, b \in \mathbf{Z}\}$. This will allow us to simplify statements about integers by covering both the cases of even and odd discriminants. For any quadratic algebraic number $\alpha = (-b+e\sqrt{d})/2a$, the *conjugate* of $\alpha$, written $\bar{\alpha}$, is $\bar{\alpha} = (-b-e\sqrt{d})/2a$. The *norm* of $\alpha$, written $N(\alpha)$, is $N(\alpha) = \alpha\bar{\alpha} = (b^2+e^2d)/4a^2$. The norm of an algebraic integer is always a rational integer and is a multiplicative function; that is, for any integers $\alpha$ and $\beta$, we have $N(\alpha\beta) = N(\alpha) \cdot N(\beta)$. An element $\varepsilon$ in a ring of quadratic algebraic integers is a *unit* of the ring if it has norm $\pm 1$. Two algebraic integers will be called *associates* if their quotient is a unit.

**Proposition 6.7.** *If $d \neq 1$ is a rational integer without square factors except 1, then the units of $\mathbf{Q}(\sqrt{d})$ are precisely the elements of $\mathcal{O}(\sqrt{d})$ which have inverses in the ring $\mathcal{O}(\sqrt{d})$, and are the integers:*

*a) if $d = -1, \pm 1$ and $\pm\sqrt{-1}$;*

b) *if* $d = -3, \pm 1$ *and* $(\pm 1 \pm \sqrt{-3})/2$, *with the* $\pm$ *signs being independent;*

c) *if* $d < -3, \pm 1$;

d) *if* $d > 1, \pm 1$ *and* $\pm \varepsilon^n$ *for any rational integer* $n$, *where* $\varepsilon$ *is the fundamental solution to the appropriate Pell equation (3.11) or (3.2).*

**Proof.** In the discussion of the Pell equation, these results were all proved. The only reason to repeat the results here is to phrase them in terms of units in a quadratic number field.

## 6.3   Ideals in Quadratic Fields

We now state, but do not prove, a special case of a major theorem.

**Theorem 6.8.** *If* **a** *is an ideal in a ring of integers* $\mathcal{O} = \mathcal{O}(\sqrt{d})$, *then there exist algebraic integers* $\alpha_1$ *and* $\alpha_2$ *in* $\mathcal{O}$ *such that any element of* **a** *can be written uniquely as* $\alpha_1 x + \alpha_2 y$, *with rational integers* $x$ *and* $y$.

Such a pair $< \alpha_1, \alpha_2 >$ is called a *basis* of the ideal **a**. We shall frequently write **a** $=< \alpha_1, \alpha_2 >$. The feature of this theorem on which we will rely is the uniqueness of representation of elements of **a**.

**Theorem 6.9.** *Any ideal* **a** *has a basis* $< a, b + g\delta >$, *where* $a$, $b$, *and* $g$ *are rational integers,* $a > 0$, $0 \leq b < a$, *and* $0 < g \leq a$, *and* $g$ *divides both* $a$ *and* $b$. *The basis with these properties is unique.*

**Proof.** We shall prove this theorem by means of several smaller propositions.

**Proposition 6.10.** *Any ideal* **a** *has a basis* $< a, \beta >$, *where* $a$ *is a rational integer and* $\beta$ *is a quadratic algebraic integer.*

**Proof.** The integers $\mathcal{O}$ are of the form $a + b\delta$, for $a$ and $b$ rational integers. This must, of course, be true for the basis elements, so we may write $\alpha_1 = a_1 + b_1\delta$ and $\alpha_2 = a_2 + b_2\delta$ with $a_1, a_2, b_1, b_2 \in \mathbf{Q}$. Since $\alpha_1 x + \alpha_2 y = \alpha_1(x + ky) + (\alpha_2 - k\alpha_1)y$ for any rational integer $k$, we have that $< \alpha_1, \alpha_2 - k\alpha_1 >$ (and by symmetry $< \alpha_1 - k\alpha_2, \alpha_2 >$) is an equivalent basis for **a**. This fact allows us to perform the Euclidean algorithm on $b_1$ and $b_2$. When this is finished, we will have a basis $< a, b+g\delta >$ for **a**, with rational integers $a$, $b$, and $g$, which is equivalent to the basis $< \alpha_1, \alpha_2 >$, and for which $0 \neq g = \gcd(b_1, b_2)$. We may then go one step further, subtracting multiples of $a$ from $b$, to be able to assume that $0 \leq b < a$.

**Proposition 6.11.** *Any rational integer $m$ which is an element of an ideal $\mathbf{a}$ is a multiple of $a$.*

**Proof.** Clearly, if there is a unique representation for $m$ by Theorem 6.8, then the representation $m = ax + (b + g\delta)y$, using the basis of Proposition 6.10, must also be unique. This implies that $y$ must be zero.

**Proposition 6.12.** $N(b + g\delta)$ *is a multiple of $a$.*

**Proof.** This norm is a rational integer and is also an element of $\mathbf{a}$; we apply Proposition 6.11.

**Proposition 6.13.** *An ideal $\mathbf{a}$ has a unique basis of the form $< a, b + g\delta >$, with rational integers $a$, $b$, and $g$, $a > 0$, $0 \leq b < a$, and $0 < g \leq a$.*

**Proof.** Given any basis for $\mathbf{a}$, we have by Proposition 6.9 an equivalent basis of the form $< a, b + g\delta >$, with $a$, $b$, and $g$ rational integers, and $a$ and $b$ satisfying the desired conditions. By Proposition 6.11, the rational integer $a$ is uniquely determined for the entire ideal, regardless of the initial choice of basis. Now, if there were distinct bases $< a, b + g\delta >$ and $< a, b' + g'\delta >$, then we could find rational integers $x$ and $y$ such that $ax + (b + g\delta)y = b' + g'\delta$. This would require us to have $ax + by = b'$ and $gy = g'$. However, if both $b$ and $b'$ lie between 0 and $a$, we must have $x = 0$ and $y = 1$. This proves that $g$ is unique. That $g$ must also satisfy the above condition follows from the following argument. Since $a$ is an element of $\mathbf{a}$, so is $a\delta$. Thus there is a unique representation $a\delta = ax + (b + g\delta)y$. This, however, implies that $gy = a$ so that we must have $0 < g \leq a$.

**Proposition 6.14.** *$g$ divides $a$ and $b$.*

**Proof.** Using Proposition 6.12 we see that any common factor of $a$ and $g$ must also divide $b$. Thus, with no loss of generality, we may assume that $a$ and $g$ are relatively prime, having already removed the common factors. Let $x$ and $y$ be rational integers chosen so that $ax + by = 1$. Since we have $a \in \mathbf{a}$ and since $\delta \in \mathcal{O}$, we have $a\delta \in \mathbf{a}$. Thus $ax\delta + (b + g\delta)y = by + \delta$ is an element of $\mathbf{a}$. However, we have a basis for $\mathbf{a}$, so rational integers $r$ and $s$ exist such that $ar + s(b+g\delta) = by + \delta$. We conclude that $sg = 1$ and that $g$, being positive, is 1.

Propositions 6.10 through 6.14 prove Theorem 6.9. The basis defined by Theorem 6.9 is called the *canonical basis* for the ideal $\mathbf{a}$. We now turn Theorem 6.9 around.

**Theorem 6.15.** *Let $a$, $b$, $g$, and $k$ be rational integers, with $a > 0$, $0 \leq b < a$, $0 < g \leq a$, and $N(b + g\delta) = ka$. Then the ideal*

$$\mathbf{a} = \{a\alpha + (b + g\delta)\beta : \alpha, \beta \in \mathcal{O}\}$$

*has canonical basis $< a, b + g\delta >$.*

**Proof.** Without loss of generality we may assume that $g = 1$. The ideal $\mathbf{a}$ has some canonical basis $< t, r + s\delta >$. Since $b + \delta$ is in $\mathbf{a}$ and since the canonical basis provides a unique representation for any element of the ideal, it is evident that we must have $s = 1$. What we shall now show is that all rational integers which are elements of $\mathbf{a}$ are multiples of $a$. This would imply that $t = a$. Since this is true, and since $b+\delta$ and $r+\delta$ are both elements of $\mathbf{a}$, so is their difference $b-r$, which is a rational integer and therefore a multiple of $a$. But from $0 \le b < a$ and $0 \le r < a$ we must then conclude that $b = r$, which proves the theorem.

Now, the elements of $\mathbf{a}$ are the numbers of the form $a\alpha + (b+\delta)\beta$, where $\alpha$ and $\beta$ are arbitrary algebraic integers of $\mathcal{O}$. That is to say, they are the numbers

$$a(x_1 + y_1\delta) + (b+\delta)(x_2 + y_2\delta)$$
$$= ax_1 + bx_2 + y_2\delta^2 + \delta(ay_1 + by_2 + x_2),$$

where $x_1$, $x_2$, $y_1$, and $y_2$ run through all the rational integers. In order that these numbers be rational integers, we must have $ay_1 + by_2 + x_2 = 0$, that is to say, $ay_1 + by_2 = -x_2$. In the case that $\delta = -\sqrt{d}$, the rational integers in $\mathbf{a}$ are thus all of the form

$$ax_1 + b(-ay_1 - by_2) + y_2 d$$
$$= ax_1 - aby_1 - (b^2 - d)y_2$$
$$= ax_1 - aby_1 - kay_2.$$

The case for odd discriminants is similar, but again more tedious.

Given any ideal $\mathbf{a} = <\alpha_1, \alpha_2>$, we define the *norm* of $\mathbf{a}$, written $N(\mathbf{a})$, to be $|\alpha_1\bar\alpha_2 - \bar\alpha_1\alpha_2| / \sqrt{\Delta}$, where $\Delta$ is the discriminant of the field of radicand $d$. None of the changes of basis used in proving Theorem 6.9 affect this quantity, so the norm of an ideal is well-defined regardless of basis and is in fact equal to $a$ if the ideal is written with the canonical basis.

The *product* of two ideals $\mathbf{a}$ with basis $<\alpha_1, \alpha_2>$ and $\mathbf{b}$ with basis $<\beta_1, \beta_2>$ is defined simply as the ideal generated by $\mathbf{Z}$-linear combinations of $\alpha_1\beta_1$, $\alpha_1\beta_2$, $\alpha_2\beta_1$, and $\alpha_2\beta_2$. A *fractional ideal* is a subset $I$ of $\mathbf{Q}(\sqrt{\Delta})$ for which the following two properties hold:

a) for any $\alpha, \beta \in I$, then for any $\lambda, \mu \in \mathcal{O}$, $\lambda\alpha + \mu\beta \in I$.

b) there exists an fixed algebraic integer $\nu$ such that, for every $\alpha \in I$, $\nu\alpha \in \mathcal{O}$.

Clearly, every ideal (hereafter called "integral ideal") is also a fractional ideal. Intuitively, the fractional ideals are the closed linear combinations with a "common denominator." If we define $\nu$ as in b), then the set $\{\nu\alpha : \alpha \in I\}$ is an integral ideal, which has a canonical basis $< a, b + g\delta >$, and thus $I$ has a basis $< a/\nu, (b + g\delta)/\nu >$.

We define an ideal **a** to be *principal* if there exists an algebraic integer $\alpha$ such that $\mathbf{a} = \{\lambda\alpha : \lambda \in \mathcal{O}\}$. A principal ideal has a canonical basis, but we shall usually write simply $(\alpha)$, the ideal being in fact the set $(\alpha)\mathcal{O}$. We define two (integral or fractional) ideals **a** and **b** to be *equivalent* if there is a principal ideal $(\alpha)$ such that $\mathbf{a} = (\alpha)\mathbf{b}$. These ideals are *narrowly equivalent* if the norm of $\alpha$ is positive. The ideal $\bar{\mathbf{a}}$ *conjugate* to **a** is the ideal with basis $< a, b + g\bar{\delta} >$. Clearly $\bar{\mathbf{a}}$ is an ideal, although this basis is not necessarily the canonical basis.

We now prove a major theorem.

**Theorem 6.16.** *Given any integral ideal* **a**, *there exists an integral ideal* **b** *such that* **ab** *is a principal ideal.*

**Proof.** This theorem is actually true for ideals in any ring of algebraic integers, although the proof in general can be much more difficult. We shall take advantage of the fact that we work only in quadratic fields to prove this in a brutal, but elementary, way. We let $\mathbf{a} = < a, b + g\delta >$ with the canonical basis. It is easy to see that we may remove the principal ideal factor $(g)$ from $\mathbf{a}$: $\mathbf{a} = (g)\mathbf{a}'$, with $\mathbf{a}'$ having a canonical basis $< a/g, (b/g + \delta) >$, and these are all integral ideals. Therefore, we need only consider ideals whose canonical basis has $g = 1$.

Consider now the product $I = \mathbf{a}\bar{\mathbf{a}}$. This is the ideal consisting of all elements of $\mathcal{O}$ of the form

$$a^2 xz + a(b + \delta)yz + a(b + \bar{\delta})xw + (b + \delta)(b + \bar{\delta})yw,$$

where $x$, $y$, $z$, and $w$ independently run through all possible rational integers. We note that $(b + \delta)(b + \bar{\delta})$ is a rational integer and is a multiple $ka$ of $a$. Thus the product consists of all elements of $\mathcal{O}$ of the form

$$a \left[ axz + (b + \delta)yz + (b + \bar{\delta})xw + kyw \right]. \tag{6.4}$$

This product ideal is, in fact, $(a)$. The principal ideal $(a)$ has canonical basis $< a, a\delta >$. We first show that the expression in brackets in (6.4) represents 1. We shall prove this only for the case $\delta = -\sqrt{d}$; the other case is similar, but more tedious and no more enlightening. It is easy to see that the expression represents $a$, $k$, and $2b$ and hence represents $\gcd(a, k, 2b)$. Now, using the equation $b^2 - d = ka$ as a congruence modulo 4, we can readily see that at least one of $a$ and $k$ must be odd so that $\gcd(a, k, 2b) = \gcd(a, k, b)$; this rational integer is represented by the expression in question. But by the same equation, if any prime $p$ divides $a$, $k$, and $b$ simultaneously, we must have $p^2$ dividing $d$; this contradicts our definition of $d$.

Now, since the expression above represents 1, it must be the case that $a$ is the smallest positive rational integer in the ideal $I$. Thus the canonical basis for $I$ is of the form $< a, t + r\sqrt{d} >$, for some rational integers $t$ and $r$. But this would imply that, for any rational integer $y$,

$(t + r\sqrt{d})y \in I$. Looking again at (6.4), we see that every element of $I$ is of the form $an + am\sqrt{d}$, for rational integers $n$ and $m$. We conclude that $t$ is zero and that $r$ is $a$, proving the theorem.

It is immediately seen that Theorem 6.16 carries over to fractional ideals as well as integral ideals. We state, without proof, a straightforward proposition and use it to derive a main theorem.

**Proposition 6.17.**

a) *The equivalence defined on ideals is an equivalence relation.*

b) *The narrow equivalence defined on ideals is an equivalence relation.*

c) *If, for ideals* **a** *and* **b**, *we have* **a** $\sim$ **b**, *then for any ideal* **c**, *we have* **ac** $\sim$ **bc**.

d) *If, for ideals* **a**, **b**, *and* **c**, *with* **c** *not the zero ideal, we have* **ac** $\sim$ **bc**, *then we have* **a** $\sim$ **b**.

**Theorem 6.18.** *The equivalence classes (respectively, narrow equivalence classes) of fractional ideals of a ring of quadratic algebraic integers $\mathcal{O}$ form an abelian group under multiplication of ideals. The identity of the group is the class of all principal ideals (respectively, the class of all principal ideals $(\alpha)$ with $N(\alpha) > 0$).*

**Proof.** The proof is immediate from part d) of the proposition above. The set of ideals by themselves forms a monoid–a set with a closed associative multiplication and an identity–and the introduction of the equivalence relation, together with Theorem 6.16, provides the means for obtaining the inverse of any class.

The groups of classes of ideals and of narrow classes of ideals are called the *class group* and the *narrow class group* of the field. The difference between equivalence and narrow equivalence is summed up as follows. For negative field discriminants $\Delta$, all norms are positive, and thus equivalence and narrow equivalence are identical. For positive field discriminants $\Delta$, norms may be positive or negative. Since the units of the ring of algebraic integers are precisely the elements of the ring which have inverses in the ring, multiplication of any ideal by a unit leaves the ideal unchanged. If there exists a unit $\varepsilon$ in $\mathcal{O}$ of norm $-1$, then all equivalent ideals are narrowly equivalent–the principal ideals $(\alpha)$ and $(\varepsilon\alpha)$ are equal, and one of $N(\alpha)$ and $N(\varepsilon\alpha)$ is positive. If there exists no such unit, then every equivalence class splits into two narrow equivalence classes.

**Theorem 6.19.**

a) *If the discriminant $\Delta$ of the quadratic field $\mathbf{Q}(\sqrt{\Delta})$ is negative, then the class group and the narrow class group are isomorphic.*

b) *If the discriminant $\Delta$ of the quadratic field $\mathbf{Q}(\sqrt{\Delta})$ is positive, and a solution exists to the equation*

$$x^2 - \Delta y^2 = -4,$$

*then the class group and the narrow class group are isomorphic.*

c) *If the discriminant $\Delta$ of the quadratic field $\mathbf{Q}(\sqrt{\Delta})$ is positive, and no solution exists to the equation*

$$x^2 - \Delta y^2 = -4,$$

*then the class group consists of the subgroup of squares of the narrow class group.*

# 6.4 Binary Quadratic Forms and Classes of Ideals

We can now prove the theorem which is the major object of this chapter.

**Theorem 6.20.** *The group of classes of binary quadratic forms of discriminant $\Delta$ is isomorphic to the narrow class group of the quadratic number field $\mathbf{Q}(\sqrt{\Delta})$.*

**Proof.** We follow somewhat the proof used in Hecke [HECK]. We shall show that to any ideal there corresponds a form, that to any form corresponds an ideal, and then that equivalent ideals correspond to equivalent forms, and conversely.

Any ideal $\mathbf{a}$ has a basis $< \alpha_1, \alpha_2 >$. We choose to order the basis so that $\alpha_1 \bar{\alpha}_2 - \bar{\alpha}_1 \alpha_2 = N(\mathbf{a})\sqrt{(\Delta)}$ is positive or positive imaginary. The binary quadratic form of discriminant $\Delta$ which we associate with this ideal is

$$\frac{[\alpha_1 x + \alpha_2 y][\bar{\alpha}_1 x + \bar{\alpha}_2 y]}{\mid N(\mathbf{a}) \mid}.$$

This is indeed a binary quadratic form with rational integer coefficients and discriminant $\Delta$, as can be verified by direct calculation. Further, for a negative discriminant, this is a positive definite form. We say that the form *belongs* to $\mathbf{a}$.

Conversely, with any binary quadratic form $(A, B, C)$ of discriminant $\Delta$ we can associate the ideal

$$\{A\alpha + (b + \delta)\beta : \alpha, \beta \in \mathcal{O}\}$$

if $A > 0$ and

$$\{A\delta\alpha + (b + \delta)\delta\beta : \alpha, \beta \in \mathcal{O}\}$$

if $A < 0$, where $b$ is $B/2$ or $(B - 1)/2$ according as $\Delta$ is even or odd.

Now, by adding or subtracting rational integral multiples of $A$ from $b$, we produce an identical (not just equivalent) ideal

$$\{A\alpha + (b' + \delta)\beta : \alpha, \beta \in \mathcal{O}\}$$

or

$$\{A\alpha\delta + (b' + \delta)\delta\beta : \alpha, \beta \in \mathcal{O}\},$$

with $0 \le b' <| A |$. Thus $<| A |, b'+\delta >$ is a canonical basis for an ideal, so that the ideal associated with the binary quadratic form $(A, B, C)$ has a basis $< A, b + \delta >$ or $< A\delta, b\delta + \delta^2 >$, depending on whether $A$ is positive or negative. Significantly, the norm of either ideal is positive or positive imaginary, and the form which belongs to that ideal is the original form $(A, B, C)$. Thus, to every binary quadratic form there corresponds an ideal to which that form belongs.

To show that equivalent forms belong to equivalent ideals and conversely, we need only show that this is true for the generators $S$ and $T$ of the modular group $\Gamma$. For the generator $S$ this has already been done–this is adding or subtracting multiples of $A$ to $b$. The other matrix produces the equivalence of forms $(A, B, C) \sim (C, -B, A)$. This corresponds to the following narrow ideal equivalence. The ideal

$$\{A\alpha + (b + \delta)\beta : \alpha, \beta \in \mathcal{O}\}$$

is narrowly equivalent to the ideal

$$\{A(b + \bar{\delta})\alpha + (b + \delta)(b + \bar{\delta})\beta : \alpha, \beta \in \mathcal{O}\}$$

if $N(b + \delta)$ is positive. If this norm is positive, then $A$ and $C$ are of like sign. We remove either the principal ideal $(A)$ or $(-A)$ to obtain an equivalent ideal

$$\{(b + \bar{\delta})\alpha + C\beta : \alpha, \beta \in \mathcal{O}\}$$

or

$$\{(-b - \delta)\alpha - C\beta : \alpha, \beta \in \mathcal{O}\}.$$

Both of these are equal to an ideal

$$\{C\alpha - (b' + \delta)\beta : \alpha, \beta \in \mathcal{O}\}.$$

If $C$ is positive, this is the ideal associated with a form $(C, B', *)$ equivalent under some matrix transformation $S^n$ to $(C, -B, A)$. If $C$ is negative, we must carry a factor of $\delta$ throughout; the form correspondence, however, is the same.

We make a final observation concerning the ability to use forms to explicitly compute in quadratic number fields. Given a field discriminant $\Delta$ and the field $\mathbf{Q}(\sqrt{\Delta})$, we can separate the rational primes $p$ into three categories.

a) $p$ divides $\Delta$. Then $p$ is representable by an ambiguous form $(p, p, *)$ of discriminant $\Delta$. Thus there is an ideal $\mathbf{p} =< p, (-p + \sqrt{\Delta})/2 >$ of norm $p$ in $\mathcal{O}$, and the ideal $\mathbf{p}$ is its own conjugate in the ring of algebraic integers. Such prime ideals $(p)$ in the rational integers are said to *ramify* in $\mathbf{Q}(\sqrt{\Delta})$. We note that the statement that the ideal is self-conjugate is equivalent to the statement that its square is a principal ideal since the product of any ideal $\mathbf{a}$ and its conjugate $\bar{\mathbf{a}}$ is always the principal ideal $(N(\mathbf{a}))$.

b. $p$ does not divide $\Delta$, and $b^2 \equiv \Delta$ (mod $p$) is solvable. Then there exist two inequivalent forms $(p, \pm b, *)$ of discriminant $\Delta$ which represent $p$. There are correspondingly two inequivalent but conjugate ideals $\mathbf{p} =< p, (-p + \sqrt{\Delta})/2 >$ and $\bar{\mathbf{p}} =< p, (+p + \sqrt{\Delta})/2 >$ of norm $p$ in $\mathcal{O}$. Their product is the principal ideal $(p)$ of norm $p^2$. In this case $p$ is said to *split* in $\mathbf{Q}(\sqrt{\Delta})$.

c) $p$ does not divide $\Delta$, and $b^2 \equiv \Delta$ (mod $p$) is not solvable. There exist no forms of discriminant $\Delta$ which represent $p$, and no ideals of norm $p$ in $\mathcal{O}$. We can always represent $p^2$, however, by the principal form if by no other way. Such primes are said to *remain prime* in $\mathbf{Q}(\sqrt{\Delta})$, and any forms which represent $p^2$ must be in the principal genus since all quadratic characters for a perfect square must be $+1$.

## 6.5 History

We have defined binary quadratic forms

$$f(x, y) = ax^2 + bxy + cy^2$$

of discriminant $b^2 - 4ac = \Delta$. This is the definition of forms according to Eisenstein, and turns out, as we have seen, to be the appropriate definition for making the correspondence between the groups of classes of forms and the classes of ideals of quadratic number fields. It is entirely for this reason that Eisenstein forms have been conisidered in this book.

Following most mathematicians before him, however, Gauss in the *Disquisitiones Arithmeticae* defined forms to be

$$\phi(x, y) = ax^2 + 2bxy + cy^2$$

of *determinant* $b^2 - ac = \delta$. There are annoying complications with either definition. With the Eisenstein forms, 2 has a special place and the odd and even discriminants usually have to be handled separately. With the forms of Gauss, both *properly primitive* forms, for which $\gcd(a, 2b, c) = 1$, and *improperly primitive* forms, for which $\gcd(a, 2b, c) = 2$, must be considered. Further complications arise, but none so serious as the fact that the groups of classes of Gauss forms do not correspond directly with all the class groups of quadratic number fields, and so do not provide the significant advantage of an elementary and explicit way for doing number-theoretic computations in those fields.

# Chapter 7

# Composition of Forms

## 7.1 Nonfundamental Discriminants

We are now ready to compare forms and groups of discriminants $\Delta$ and $\Delta r^2$. In the language of algebraic number theory, this is a comparison of the group of classes of ideals in the ring of integers with the group of classes of ideals in the *order* of index $r$. We recall that if

$$R = \begin{pmatrix} \alpha & \beta \\ \gamma & \delta \end{pmatrix}$$

is any $2 \times 2$ matrix with integer coefficients and determinant $r$, then the change of variables (1.1) takes a form $f = (a,\, b,\, c)$ of discriminant $\Delta$ to a form

$$f' = (a',\, b',\, c') = (a\alpha^2+b\alpha\gamma+c\gamma^2, b(\alpha\delta+\beta\gamma)+2(a\alpha\beta+c\gamma\delta), a\beta^2+b\beta\delta+c\delta^2)$$

of discriminant $\Delta r^2$. In matrix notation this is

$$\begin{pmatrix} a' & b'/2 \\ b'/2 & c \end{pmatrix} = \begin{pmatrix} \alpha & \gamma \\ \beta & \delta \end{pmatrix} \begin{pmatrix} a & b/2 \\ b/2 & c \end{pmatrix} \begin{pmatrix} \alpha & \beta \\ \gamma & \delta \end{pmatrix},$$

which we will write as $f' = R^T f R$ for brevity. We shall call such a matrix $R$ a *transformation of determinant* $r$ and shall say that $f'$ is *derived* from $f$ by the transformation of determinant $r$.

**Proposition 7.1.**

a) *Given any primitive form $f$ of discriminant $\Delta r^2$, there exists a form $f'$ of discriminant $\Delta$ and a transformation $R$ of determinant $r$ such that $f = R^T f' R$.*

a) *If $f$ and $f'$ are primitive equivalent forms of discriminant $\Delta r^2$, then there exists a primitive form $f''$ of discriminant $\Delta$ and transformations $R$ and $R'$ of determinant $r$ and a matrix $M \in \Gamma$ such that*

$$
\begin{aligned}
f &= R^T f'' R \\
f' &= (R')^T f''(R') \\
R' &= RM.
\end{aligned}
$$

**Proof.** a) Given $f$, there exists, under a transformation $M \in \Gamma$, an equivalent form $(a', b'r, c'r^2)$, in which $a'$ is relatively prime to $r$. We write $f' = (a', b', c')$ and

$$
R = \begin{pmatrix} 1 & 0 \\ 0 & r \end{pmatrix} M.
$$

Then clearly $f = R^T f' R$.

b) Given $f$, there exists $N_1 \in \Gamma$ such that $N_1^T f N_1$ is the form $(a', b'r, c'r^2)$, with $a'$ relatively prime to $r$. There must also exist an $N_2 \in \Gamma$ such that $N_2^T f' N_2$ is the same form. Let $f'' = (a', b', c')$. Then

$$
N_1^T f N_1 = N_2^T f' N_2 = \begin{pmatrix} 1 & 0 \\ 0 & r \end{pmatrix} f'' \begin{pmatrix} 1 & 0 \\ 0 & r \end{pmatrix}.
$$

We let

$$R = \begin{pmatrix} 1 & 0 \\ 0 & r \end{pmatrix} N_1^{-1},$$

$$R' = \begin{pmatrix} 1 & 0 \\ 0 & r \end{pmatrix} N_2^{-1},$$

and $M = N_1 N_2^{-1}$.

The meaning of Proposition 7.1 is that any primitive form of discriminant $\Delta r^2$ can be derived from a primitive form of discriminant $\Delta$ by the application of a transformation $R$ of determinant $r$, and that equivalent forms of discriminant $\Delta r^2$ can only be derived from the same form of discriminant $\Delta$. It remains to be seen how many inequivalent forms can be derived from the same form.

We restrict ourselves to the case where $r$ is a prime $p$. It will turn out that the general case can be built from this case. We define two transformations $R_1$ and $R_2$ of determinant $p$ to be $\Gamma$-*right-equivalent* if there exists some $M \in \Gamma$ such that $R_1 M = R_2$. This is easily seen to be an equivalence relation. If $f$ is a primitive form of discriminant $\Delta$, and $f'$ and $f''$ are forms of discriminant $\Delta p^2$ derived by $\Gamma$-right-equivalent matrices from $f$, then $f'$ and $f''$ are equivalent in the ordinary sense as binary quadratic forms.

**Proposition 7.2.** *The $\Gamma$-right-equivalent transformations have as equivalence class representatives the $p + 1$ transformations*

$$\begin{pmatrix} 1 & 0 \\ 0 & p \end{pmatrix}$$

*and*

$$\begin{pmatrix} p & h \\ 0 & 1 \end{pmatrix},$$

*for* $0 \le h \le p - 1.$

**Proof.** Let

$$R = \begin{pmatrix} \alpha & \beta \\ \gamma & \delta \end{pmatrix}$$

be any transformation of determinant $p$. Since $\alpha\delta - \beta\gamma = p$, the greatest common divisor of $\gamma$ and $\delta$ must be either 1 or $p$.

*Case 1.* $\gcd(\gamma, \delta) = 1$. If we choose $\beta'$ and $\delta'$ such that $\beta'\gamma + \delta\delta' = 1$, then

$$\begin{pmatrix} \alpha & \beta \\ \gamma & \delta \end{pmatrix} \begin{pmatrix} \delta & \beta' \\ -\gamma & \delta' \end{pmatrix} = \begin{pmatrix} \alpha\delta - \beta\gamma & \alpha\beta' + \beta\delta' \\ \gamma\delta - \gamma\delta & \beta'\gamma + \delta\delta' \end{pmatrix} = \begin{pmatrix} p & pk + h \\ 0 & 1 \end{pmatrix}$$

for some $0 \le h \le p - 1$. By further multiplication on the right by

$$\begin{pmatrix} 1 & -k \\ 0 & 1 \end{pmatrix}$$

we get

$$\begin{pmatrix} p & h \\ 0 & 1 \end{pmatrix}.$$

*Case 2.* $\gcd(\gamma, \delta) = p$. Then $\gamma = p\gamma''$ and $\delta = p\delta''$, and we can choose $\beta'$ and $\delta'$ such that $\beta'\gamma'' + \delta'\delta'' = 1$. Then

$$\begin{pmatrix} \alpha & \beta \\ \gamma & \delta \end{pmatrix} \begin{pmatrix} \delta'' & \beta' \\ -\gamma'' & \delta' \end{pmatrix} = \begin{pmatrix} 1 & h \\ 0 & p \end{pmatrix}.$$

We multiply on the right by

$$\begin{pmatrix} 1 & -h \\ 0 & 1 \end{pmatrix}$$

to get

$$\begin{pmatrix} 1 & 0 \\ 0 & p \end{pmatrix}.$$

The proposition is proved.

Now, every primitive class of discriminant $\Delta p^2$ can be derived by applying one of the $p + 1$ $\Gamma$-right-equivalence class representatives to some form of discriminant $\Delta$, $(a, b, c)$, with $\gcd(a, p) = 1$. We must determine how many of these derived forms are primitive. We recall the definition of the Kronecker symbol $\chi_\Delta(p)$, which is the Jacobi symbol $\left(\frac{\Delta}{p}\right)$ if $p$ and $\Delta$ are relatively prime, and 0 if they are not.

**Proposition 7.3.** *Given any form* $(a, b, c)$ *of discriminant* $\Delta$, *and any prime* $p$ *with* $\gcd(a, p) = 1$, *the* $p + 1$ *representative transformations of determinant* $p$ *produce exactly* $p - \chi_\Delta(p)$ *primitive forms of discriminant* $\Delta p^2$.

**Proof.** First, consider the case of an odd prime $p$. Then, since

$$\begin{pmatrix} a & bp/2 \\ bp/2 & cp^2 \end{pmatrix} = \begin{pmatrix} 1 & 0 \\ 0 & p \end{pmatrix} \begin{pmatrix} a & b/2 \\ b/2 & c \end{pmatrix} \begin{pmatrix} 1 & 0 \\ 0 & p \end{pmatrix},$$

the form $(a, bp, cp^2)$ derived by applying the transformation

$$\begin{pmatrix} 1 & 0 \\ 0 & p \end{pmatrix}$$

is obviously primitive. It is easily seen that, under the transformation

$$\begin{pmatrix} p & h \\ 0 & 1 \end{pmatrix},$$

the form $(ap^2, p(b+2ah), ah^2 + bh + c)$ is derived from $(a, b, c)$. Since $(a, b, c)$ is primitive, the only common divisors of the coefficients of the derived form can be 1, $p$, or $p^2$. But since $\gcd(a, p) = 1$, we see that

$$ah^2 + bh + c \equiv 0 \pmod{p}$$

if and only if

$$(2ah + b)^2 \equiv \Delta \pmod{p}.$$

That is to say, the form derived from $(a,\ b,\ c)$ is primitive if and only if the congruence in $h$,

$$(2ah + b)^2 \equiv \Delta \pmod{p} \tag{7.1}$$

has no solutions.

If the discriminant $\Delta$ is divisible by $p$, then (7.1) has exactly one solution, which is the value of $h$ for which $b + 2ah$ is congruent to zero modulo $p$. If $\Delta$ is prime to $p$, and the congruence (7.1) has any solutions, then it has exactly two, modulo $p$. Thus, from any class of forms of discriminant $\Delta$, we can derive exactly $p - \chi_\Delta(p)$ primitive forms of discriminant $\Delta p^2$.

We consider the remaining case of the prime 2. The three derived forms of discriminant $4\Delta$ are $(a,\ 2b,\ 4c)$, $(4a,\ 2b,\ c)$, and $(4a,\ 2b + 4a,\ a + b + c)$. The first is always primitive. The second is primitive if and only if $c$ is odd, and the third is primitive if and only if $a + b + c$ is odd, that is, if and only if $b + c$ is even. For odd discriminants $\Delta$, $b$ is odd, so both the second and third derived forms are primitive if and only if $c$ is odd. It is easy to see that $c$ is odd if and only if $\Delta \equiv 5 \pmod{8}$.

For even discriminants $\Delta$, $b$ is always even, so the second derived form is primitive if and only if $c$ is odd, and the third derived form is primitive if and only if $c$ is even. Clearly, then, exactly one of these two is primitive. This special case fits our earlier formula $p - \chi_\Delta(p)$ for the number of primitive forms of discriminant $\Delta p^2$ derived from any given class of discriminant $\Delta$. The proposition is proved.

Now, which of these derived forms are equivalent? Let $f = (a, b, c)$ be a form of discriminant $\Delta$ and $p$ a prime such that $\gcd(a, p) = 1$; assume that transformations of determinant $p$, $R_1$ and $R_2$, exist so that $R_1^T f R_1 \sim R_2^T f R_2$. Then there exists $M \in \Gamma$ such that $R_1^T f R_1 = M^T R_2^T f R_2 M$, so $f = (R_2 M R_1^{-1})^T f (R_2 M R_1^{-1})$. Exhaustively considering cases, we see that any such matrix $R_2 M R_1^{-1}$, where $R_1$ and $R_2$ are representative transformations of the same prime determinant, is an element of $\Gamma$ and is therefore an automorph of $f$. As was shown in Chapter 3, all the automorphs are of the form

$$\begin{pmatrix} (X_i - bY_i)/2 & -cY_i \\ aY_i & (X_i + bY_i)/2 \end{pmatrix},$$

where $(X_i, Y_i)$ is an integral solution of $X^2 - \Delta Y^2 = 4$.

In order for equivalent forms to be derived from the same form $f$, then, we must have an automorph $A$ of $f$ and representative transformations $R_1$ and $R_2$ of determinant $p$ such that $R_2^{-1} A R_1$ is a matrix $M$ in the modular group $\Gamma$. We have two cases to consider:

a) If

$$R_1 = \begin{pmatrix} p & h \\ 0 & 1 \end{pmatrix}, R_2 = \begin{pmatrix} 1 & 0 \\ 0 & p \end{pmatrix},$$

then $R_2^{-1} A R_1 \in \Gamma$ if and only if

$$(X_i + bY_i)/2 + ahY_i \equiv 0 \pmod{p}. \tag{7.2}$$

b) If

$$R_1 = \begin{pmatrix} p & h \\ 0 & 1 \end{pmatrix}, R_2 = \begin{pmatrix} p & k \\ 0 & 1 \end{pmatrix},$$

then $R_2^{-1} A R_1 \in \Gamma$ if and only if

$$h(X_i - bY_i)/2 - k(X_i + bY_i)/2 - (c + akh)Y_i \equiv 0 \quad (\text{mod } p). \quad (7.3)$$

Consider the case of an odd prime $p$. We recall from Theorem 3.21 that the values of $X_i$ and $Y_i$ are periodic with some period $s$ modulo $p$, and that $Y_s$ is congruent to zero modulo $p$. If $Y_i$ is congruent to zero modulo $p$, then $X_i$ cannot be congruent to 0, so we have exactly $s - 1$ values $Y_1, \ldots, Y_{s-1}$, for which some value $h_i$ exists so that (7.2) is solvable. These $h_i$ are distinct since inverses in the group of integers modulo $p$ are unique. Thus the $s$ forms of discriminant $\Delta p^2$ derived from any given form by the representative transformations

$$\begin{pmatrix} 1 & 0 \\ 0 & p \end{pmatrix} \quad \text{and} \quad \begin{pmatrix} p & h_i \\ 0 & 1 \end{pmatrix}, i = 1, \ldots, s - 1$$

are all distinct but equivalent.

Now, solving (7.3) for $k$, we find that

$$k \equiv \frac{h(X_i - bY_i) - 2cY_i}{X_i + bY_i + 2ahY_i} \quad (7.4)$$

holds unless the denominator is zero modulo $p$, in which case the numerator must also be zero modulo $p$. But if we have

$$h(X_i - bY_i) - 2cY_i \equiv 0 \quad (\text{mod } p)$$

and

$$X_i + bY_i + 2ahY_i \equiv 0, \quad (\text{mod } p),$$

then we have two linear congruences in two unknowns, from which we eliminate $X_i$ to find that

$$(ah^2 + bh + c)Y_i \equiv 0 \quad (\text{mod } p)$$

must hold. Since the form is primitive, we conclude that $Y_i \equiv 0$ (mod $p$) and then, looking at the above equations, that $X_i \equiv 0$ (mod $p$). But $X_i$ and $Y_i$ cannot both be divisible by an odd prime $p$. Thus it cannot happen that (7.4) is indeterminate.

Given $h$, there does exist a $k$ (unique modulo $p$) such that the two representative transformations produce equivalent forms. More precisely, given $h$ and a pair $(X_i, Y_i)$, there exists a unique $k_i$ modulo $p$ such that the corresponding derived forms are equivalent. Given $h$ and considering any pairs $(X_i, Y_i)$, and $(X_j, Y_j)$, we find that we must have

$$k_i - k_j \equiv 2(ah^2 + bh + c)(X_i Y_j - X_j Y_i) \quad (\text{mod } p).$$

On the other hand,

$$
\begin{aligned}
\frac{X_{j-i} + \sqrt{\Delta} Y_{j-i}}{2} &= \left(\frac{X_1 + \sqrt{\Delta} Y_1}{2}\right)^j \cdot \left(\frac{X_1 + \sqrt{\Delta} Y_1}{2}\right)^{-i} \\
&= \left(\frac{X_j + \sqrt{\Delta} Y_j}{2}\right) \cdot \left(\frac{X_1 - \sqrt{\Delta} Y_1}{2}\right)^i \\
&= \left(\frac{X_j + \sqrt{\Delta} Y_j}{2}\right) \cdot \left(\frac{X_i - \sqrt{\Delta} Y_i}{2}\right) \\
&= \frac{X_i X_j - \Delta Y_i Y_j + \sqrt{\Delta}(X_i Y_j - X_j Y_i)}{4}
\end{aligned}
$$

We see that $k_j - k_i$ depends on the congruence class modulo $p$ of $Y_{j-i}$ so that there are, as with (7.2), $s$ equivalent forms derived from the same form. We have proved the following theorem.

**Theorem 7.4.** *Let $p$ be an odd prime, $h(\Delta)$ and $h(\Delta p^2)$ the order of the class groups of discriminants $\Delta$ and $\Delta p^2$, respectively, and let $s$ be the least subscript $i$ such that $Y_i \equiv 0$ (mod $p$), where the pairs $(X_i, Y_i)$*

*are the solutions to the equation* $x^2 - \Delta y^2 = 4$. *Then*

$$h(\Delta p^2) = \frac{h(\Delta) \cdot (p - \chi_\Delta(p))}{s}.$$

In the above arguments we use several times the fact that $p$ is not 2. The case $p = 2$ is entirely similar, although we must take greater care because denominators and coefficients happen to be 2 in several instances. It is straightforward, though tedious, to prove the following theorem.

**Theorem 7.5.** *Let* $h(\Delta)$ *and* $h(4\Delta)$ *be the order of the class groups of discriminants* $\Delta$ *and* $4\Delta$, *respectively, and let* $(X_1, Y_1)$ *be the fundamental solution of* $x^2 - \Delta y^2 = 4$. *Then*

   *a) if* $\Delta \equiv 1$ (mod 8), *then* $h(4\Delta) = h(\Delta)$;

   *b) if* $\Delta \equiv 5$ (mod 8) *and* $Y_1$ *is even, then* $h(4\Delta) = 3h(\Delta)$;

   *c) if* $\Delta \equiv 5$ (mod 8) *and* $Y_1$ *is odd, then* $h(4\Delta) = h(\Delta)$;

   *d) if* $\Delta \equiv 0$ (mod 4) *and* $Y_1$ *is even, then* $h(4\Delta) = 2h(\Delta)$;

   *e) if* $\Delta \equiv 0$ (mod 4) *and* $Y_1$ *is odd, then* $h(4\Delta) = h(\Delta)$.

We can now finish the discussion of genera by generalizing Theorem 4.20.

**Theorem 7.6.** *Let* $\Delta$ *be any discriminant. Then exactly half the possible genera actually exist.*

**Proof.** The theorem has already been proved if $\Delta$ is fundamental. Given $\Delta$, there is some fundamental discriminant $D$ and an integer

$m$ such that $\Delta = Dm^2$. Exactly half the possible genera exist for discriminant $D$. Let $p$ be an odd prime dividing $m$. If $p$ divides $D$, then no new genera are created in constructing the forms of discriminant $Dm^2$ from those of discriminant $D$, and the theorem is true. If $p$ does not divide $D$, then one new generic character is added, doubling the number of possible genera. We must show that each of these exists.

By Proposition 4.2 we may choose as equivalence class representatives for forms of discriminant $D$ forms $(a,\ b,\ c)$ for which $p$ does not divide $a$. Then, by Proposition 7.1, all the classes of forms of discriminant $Dm^2$ are to be found among the classes of forms of $(a,\ bp,\ cp^2)$, and the assigned generic characters which originally existed for discriminant $D$ are unchanged for discriminant $Dp^2$. Again appealing to Dirichlet's theorem, we see that primes must exist in the correct congruence classes to guarantee that each genus of discriminant $D$ actually does split into two genera of discriminant $Dp^2$.

The proof for the prime $p = 2$ is similar, and the extension from discriminant $Dp^2$ to discriminant $Dp^2p'^2$ should be apparent.

Before leaving this section, we restate the formulas: All the primitive classes of discriminant $\Delta p^2$ for a prime $p$ are represented by the classes of the primitive forms in the list:

$$(a,\ bp,\ cp^2), \tag{7.5}$$
$$(ap^2,\ p(b + 2ah),\ ah^2 + bh + c),$$

where $h$ runs through all values from $0$ to $p - 1$.

## 7.2   The General Problem of Composition

We have defined composition, for forms of the same discriminant, by describing a composition algorithm which produces a group structure on the classes of forms. Before presenting the analogous algorithm for forms of different discriminants, we shall describe the general technique of composition so that at the end it will be clear that the composition algorithm used is, in fact, the unique algorithm with the desired properties. We shall follow closely the development in Mathews, which itself follows the presentation in H. J. S. Smith's *Report* [SMIT].

To this end, let us suppose that we have a form

$$F = AX^2 + BXY + CY^2$$

which is not necessarily primitive, and has discriminant $D$ and *divisor* $\gcd(A, B, C) = M$. What we seek are conditions for the existence of a transformation

$$\begin{pmatrix} X \\ Y \end{pmatrix} = \begin{pmatrix} p_0 & p_1 & p_2 & p_3 \\ q_0 & q_1 & q_2 & q_3 \end{pmatrix} \begin{pmatrix} x_1 x_2 \\ x_1 y_2 \\ y_1 x_2 \\ y_1 y_2 \end{pmatrix}, \qquad (7.6)$$

with $p_i$ and $q_i$ all integral, such that the form $F$ becomes $f_1 \cdot f_2$, where $f_1 = a_1 x_1^2 + B x_1 y_1 + a_2 C y_1^2$ of discriminant $d_1$ and divisor $m_1$ and $f_2 = a_2 x_2^2 + B x_2 y_2 + a_1 C y_2^2$ of discriminant $d_2$ and divisor $m_2$. We let

$$
\begin{aligned}
P_{01} &= p_0 q_1 - p_1 q_0 \\
P_{02} &= p_0 q_2 - p_2 q_0 \\
P_{03} &= p_0 q_3 - p_3 q_0
\end{aligned}
$$

$$P_{12} = p_1 q_2 - p_2 q_1$$
$$P_{13} = p_1 q_3 - p_3 q_1$$
$$P_{23} = p_2 q_3 - p_3 q_2$$

and we observe that

$$P_{01} P_{23} - P_{02} P_{13} + P_{03} P_{12} = 0.$$

Now the transformation (7.6) is also a transformation

$$\begin{pmatrix} X \\ Y \end{pmatrix} = \begin{pmatrix} p_0 x_2 + p_1 y_2 & p_2 x_2 + p_3 y_2 \\ q_0 x_2 + q_1 y_2 & q_2 x_2 + q_3 y_2 \end{pmatrix} \begin{pmatrix} x_1 \\ y_1 \end{pmatrix},$$

which converts $(A,\ B,\ C)$ into $(f_2 a_1,\ f_2 b_1,\ f_2 c_1)$. This transformation has determinant

$$\Delta_2 = P_{02} x_2^2 + (P_{03} + P_{12}) x_2 y_2 + P_{13} y_2^2.$$

Comparing discriminants, then, we have

$$f_2^2 \cdot d_1 = D \cdot \Delta_2^2.$$

Similarly, under a transformation of determinant

$$\Delta_1 = P_{01} x_1^2 + (P_{03} - P_{12}) x_1 y_1 + P_{23} y_1^2,$$

we convert $(A,\ B,\ C)$ into $(f_1 a_2,\ f_1 b_2,\ f_1 c_2)$, and see that

$$f_1^2 \cdot d_2 = D \cdot \Delta_1^2.$$

Let $\delta_1 = \gcd(P_{01}, P_{03} - P_{12}, P_{23})$ and $\delta_2 = \gcd(P_{02}, P_{03} + P_{12}, P_{13})$. Then we have

$$m_2^2 \cdot d_1 = D\delta_2^2$$
$$m_1^2 \cdot d_2 = D\delta_1^2.$$

We reach the significant conclusion that it is necessary that $d_1/D$ and $d_2/D$ be rational squares $n_1^2$ and $n_2^2$, respectively.

If we let $u = \gcd(\delta_1, \delta_2)$ and $v = \gcd(P_{01}, P_{02}, P_{03}, P_{12}, P_{13}, P_{23})$, then clearly we have $u = v$ or $u = 2v$. If the latter holds, then considering the equation

$$\frac{P_{01}P_{23}}{v^2} - \frac{P_{02}P_{13}}{v^2} + \frac{P_{03}P_{12}}{v^2} = 0,$$

we would have that the first two terms are even and the third term odd, which is impossible. Hence $u = v$ and

$$\gcd(d_1 m_2^2, d_2 m_1^2) = Dv^2.$$

Now we have that $\Delta_2^2 = n_1^2 f_2^2$, and similarly for the other equation, so we may choose $n_1$ and $n_2$ positive or negative so as to force

$$\Delta_1 = n_2 f_1$$
$$\Delta_2 = n_1 f_2$$

Considering each coefficient in the above equations, then we have that

$$\frac{P_{01}}{a_1} = \frac{P_{03} - P_{12}}{b_1} = \frac{P_{23}}{c_1} = n_2$$

and

$$\frac{P_{02}}{a_2} = \frac{P_{03} + P_{12}}{b_2} = \frac{P_{13}}{c_2} = n_1.$$

After Mathews, one can say that the form $f_1$ (respectively $f_2$) is taken directly or inversely according as $n_2$ (respectively $n_1$) is positive or negative. We shall soon consider only positive values and see that negative values have the effect of inverting in the class group the result of the composition.

By direct computation one can see that

$$
\begin{aligned}
\Delta_1 \Delta_2 \\
&= (q_1 q_2 - q_0 q_3) X^2 + (p_0 q_3 + p_3 q_0 - p_1 q_2 - p_2 q_2) XY + \\
&\quad (p_1 p_2 - p_0 p_3) Y^2 \\
&= n_1 f_2 n_2 f_1 = n_1 n_2 F.
\end{aligned}
$$

These are all necessary results from the existence of a suitable transformation (7.6). That the following theorem gives both necessary and sufficient conditions is entirely straightforward.

**Theorem 7.7.** *There exists a transformation*

$$
\begin{pmatrix} X \\ Y \end{pmatrix} = \begin{pmatrix} p_0 & p_1 & p_2 & p_3 \\ q_0 & q_1 & q_2 & q_3 \end{pmatrix} \begin{pmatrix} x_1 x_2 \\ x_1 y_2 \\ y_1 x_2 \\ y_1 y_2 \end{pmatrix},
$$

*with $p_i$ and $q_i$ all integral, such that the form $F$ becomes $f_1 \cdot f_2$, where*
*$f_1 = a_1 x_1^2 + B x_1 y_1 + a_2 C y_1^2$ of discriminant $d_1$ and divisor $m_1$ and*
*$f_2 = a_2 x_2^2 + B x_2 y_2 + a_1 C y_2^2$ of discriminant $d_2$ and divisor $m_2$, if and*
*only if $d_1/D = n_1{}^2$ and $d_2/D = n_2{}^2$ with rational numbers $n_1$ and $n_2$*
*and the following equations are simultaneously solvable.*

$$
\begin{aligned}
P_{01} &= p_0 q_1 - p_1 q_0 = a_1 n_2 \\
P_{02} &= p_0 q_2 - p_2 q_0 = a_2 n_1 \quad\quad (7.7)
\end{aligned}
$$

$$P_{03} + P_{12} = (p_0 q_3 - p_3 q_0) + (p_1 q_2 - p_2 q_1) = b_2 n_1$$

$$P_{03} - P_{12} = (p_0 q_3 - p_3 q_0) - (p_1 q_2 - p_2 q_1) = b_1 n_2$$

$$P_{13} = p_1 q_3 - p_3 q_1 = c_1 n_2$$

$$P_{23} = p_2 q_3 - p_3 q_2 = c_2 n_1$$

$$q_1 q_2 - q_0 q_3 = A n_1 n_2$$

$$p_0 q_3 + p_3 q_0 - p_1 q_2 - p_2 q_1 = B n_1 n_2$$

$$p_1 p_2 - p_0 p_3 = C n_1 n_2$$

Now, we have the identity

$$AX^2 + BXY + CY^2 = (a_1 x_1^2 + b_1 x_1 y_1 + c_1 y_1^2)(a_2 x_2^2 + b_2 x_2 y_2 + c_2 y_2^2).$$

Clearly we have that $M \mid m_1 m_2$. We can also show that $m_1 m_2 \mid M v^2$.
For by equating the coefficients of $x_1^2$, $x_1 y_1$, and $y_1^2$ in the above identity,
we have

$$a_1 f_2 = A \left( \frac{\partial X}{\partial x_1} \right)^2 + B \left( \frac{\partial X}{\partial x_1} \right) \left( \frac{\partial Y}{\partial x_1} \right) + C \left( \frac{\partial Y}{\partial x_1} \right)^2$$

$$b_1 f_2 = A \left( \frac{\partial X}{\partial x_1} \right) \left( \frac{\partial X}{\partial y_1} \right) + B \left( \frac{\partial X}{\partial x_1} \right) \left( \frac{\partial Y}{\partial y_1} \right) + C \left( \frac{\partial X}{\partial x_1} \right) \left( \frac{\partial Y}{\partial y_1} \right)$$

$$c_1 f_2 = A \left( \frac{\partial X}{\partial y_1} \right)^2 + B \left( \frac{\partial X}{\partial y_1} \right) \left( \frac{\partial Y}{\partial y_1} \right) + C \left( \frac{\partial Y}{\partial y_1} \right)^2$$

Multiplying and adding so as to cancel the $B$ and $C$ terms, we get

$$A \left[ \left( \frac{\partial X}{\partial x_1} \right) \left( \frac{\partial Y}{\partial y_1} \right) - \left( \frac{\partial X}{\partial y_1} \right) \left( \frac{\partial Y}{\partial x_1} \right) \right]^2 =$$

$$\left[ a \left( \frac{\partial Y}{\partial y_1} \right)^2 - b \left( \frac{\partial Y}{\partial x_1} \right) \left( \frac{\partial Y}{\partial y_1} \right) + c \left( \frac{\partial Y}{\partial x_1} \right)^2 \right] f_1.$$

The right-hand side is divisible by $m_1m_2$, and thus so must the left-hand side, which is equal to $A\Delta_2^2$. Similarly, it can be shown that $B\Delta_2^2$ and $C\Delta_2^2$ are likewise divisible. Thus $m_1m_2 \mid M\delta_2^2$. Similarly $m_1m_2 \mid M\delta_1^2$ and hence $m_1m_2 \mid Mv^2$.

We define the form $F$ to be *compounded* from $f_1$ and $f_2$ if the transformation (7.6) is primitive, that is, if $v = 1$. In this case we have

$$D \;=\; \gcd(d_1m_2{}^2, d_2m_1{}^2)$$
$$m_1m_2 \;=\; M$$

It is a straightforward matter now to see that in the case of composition, we need only look at primitive forms. Composition exists for imprimitive triples of forms if and only if it exists for the corresponding primitive forms. We may thus take $m_1 = m_2 = M = 1$. This forces $n_1$ and $n_2$ to be integers and to be relatively prime.

We have defined composition for individual forms, and must once again show that it is actually an operation on classes. Assume, then, that two transformations exist

$$\begin{pmatrix} X \\ Y \end{pmatrix} = \begin{pmatrix} p_0 & p_1 & p_2 & p_3 \\ q_0 & q_1 & q_2 & q_3 \end{pmatrix} \begin{pmatrix} x_1x_2 \\ x_1y_2 \\ y_1x_2 \\ y_1y_2 \end{pmatrix}$$

$$\begin{pmatrix} X' \\ Y' \end{pmatrix} = \begin{pmatrix} p'_0 & p'_1 & p'_2 & p'_3 \\ q'_0 & q'_1 & q'_2 & q'_3 \end{pmatrix} \begin{pmatrix} x_1x_2 \\ x_1y_2 \\ y_1x_2 \\ y_1y_2 \end{pmatrix}$$

by which the composition of $f_1$ and $f_2$ becomes, respectively, $F$ and $F'$. We must show that $F \sim F'$.

We know that the transformations are primitive and thus that integers $\lambda_{ij}$ exist by which we can write

$$\lambda_{01}P_{01} + \lambda_{02}P_{02} + \ldots + \lambda_{23}P_{23} = 1.$$

If we then let

$$\sum \lambda_{ij}(p'_i q_j - q_i p'_j) \ = \alpha$$
$$\sum \lambda_{ij}(p_i p'_j - p'_i p_j) \ = \beta$$
$$\sum \lambda_{ij}(q'_i q_j - q_i q'_j) \ = \gamma$$
$$\sum \lambda_{ij}(p_i q'_j - q'_i p_j) \ = \delta$$

where the subscripts $ij$ run in order 01, 02, 03, 12, 13, and 23, we can then show that $F$ and $F'$ are equivalent under

$$\begin{pmatrix} \alpha & \beta \\ \gamma & \delta \end{pmatrix}.$$

This is because for any fixed subscript $k$,

$$
\begin{aligned}
\alpha p_k + \beta q_k &= \sum \lambda_{ij} p_k (p'_i q_j - q_i p'_j) + \sum \lambda_{ij} q_k (p_i p'_j - p'_i p_j) \\
&= \sum \lambda_{ij} \left[ p'_i (p_k q_j - p_j q_k) - p'_j (p_k q_i - p_i q_k) \right] \\
&= \sum \lambda_{ij} \left[ p'_i (p'_k q'_j - p'_j q'_k) - p'_j (p'_k q'_i - p'_i q'_k) \right] \\
&= \sum \lambda_{ij} \left[ p'_i p'_k q'_j - p'_j p'_k q'_k \right] \\
&= \sum \lambda_{ij} p'_k (p'_i q'_j - p'_j q'_k) \\
&= p'_k \sum \lambda_{ij} (p_i q_j - p_j q_k) = p'_k.
\end{aligned}
$$

We can similarly show that $\gamma p_k + \delta q_k = q'_k$. Thus, for any $i, j$,

$$\begin{vmatrix} p'_i & p'_j \\ q'_i & q'_j \end{vmatrix} =$$

$$\begin{vmatrix} \alpha p_i + \beta q_i & \alpha p_j + \beta q_j \\ \gamma p_i + \delta q_i & \gamma p_j + \delta q_j \end{vmatrix} =$$

$$(\alpha\delta - \beta\gamma) \cdot \begin{vmatrix} p_i & p_j \\ q_i & q_j \end{vmatrix}.$$

Thus $\alpha\delta - \beta\gamma = 1$.

Now, from the transformation (7.6) considered both for $F$ and $F'$, we get that

$$X' = \alpha X + \beta Y$$
$$Y' = \gamma X + \delta Y$$

and thus that $F \sim F'$.

The converse, that if $F \sim F'$ and $F$ can be compounded of $f_1$ and $f_2$, then $F'$ can be compounded of $f_1$ and $f_2$, is obvious by reversing the calculations.

Thus the composition of forms $f_1$ and $f_2$ is not properly just a form but an entire class. We now show that the composition is not properly done just on forms but on entire classes. Assume that $F$ is compounded of $f_1$ and $f_2$ under (7.6) and that $f'_1 = (a'_1, \ b'_1, \ c'_1)$ is equivalent to $f_1$ under the transformation

$$\begin{pmatrix} x_1 \\ y_1 \end{pmatrix} = \begin{pmatrix} \alpha & \beta \\ \gamma & \delta \end{pmatrix} \begin{pmatrix} x'_1 \\ y'_1 \end{pmatrix}.$$

Then if we consider

$$\begin{pmatrix} X \\ Y \end{pmatrix}$$

$$= \begin{pmatrix} \alpha p_0 + \gamma p_2 & \alpha p_1 + \gamma p_3 & \beta p_0 + \delta p_2 & \beta p_3 + \delta p_3 \\ \alpha q_0 + \gamma q_2 & \alpha q_1 + \gamma q_3 & \beta q_0 + \delta q_2 & \beta q_3 + \delta q_3 \end{pmatrix} \begin{pmatrix} x'_1 x_2 \\ x'_1 y_2 \\ y'_1 x_2 \\ y'_1 y_2 \end{pmatrix},$$

we have the following results:

$$
\begin{aligned}
P'_{01} &= \alpha^2 P_{01} + \alpha\gamma(P_{03} - P_{12}) + \gamma^2 P_{23} \\
&= n_2(a_1\alpha^2 + b_1\alpha\gamma + c_1\gamma^2) = n_2 a'_1 \\
P'_{03} - P'_{12} &= n_2 b'_1 \\
P'_{23} &= n_2 c'_1 \\
P'_{02} &= (\alpha\delta - \beta\gamma)n_1 a_2 \\
P'_{03} + P'_{12} &= (\alpha\delta - \beta\gamma)n_1 b_2 \\
P'_{13} &= (\alpha\delta - \beta\gamma)n_1 c_2
\end{aligned}
$$

In the remaining three equations of (7.7) we can directly replace $An_1 n_2$, $Bn_1 n_2$, and $Cn_1 n_2$ by $(\alpha\delta - \beta\gamma)$ times these quantities.

We have, except in this section, always assumed that equivalence was proper equivalence, with $\alpha\delta - \beta\gamma = +1$. If this is the case, then the signs of $n'_1$ and $n'_2$ must be the same as those of $n_1$ and $n_2$ in order for the composition to take place on classes. If, however, $f_1$ and $f'_1$ are improperly equivalent, then composition of the classes can be accomplished by changing the sign of $n'_1$. Since improperly equivalent forms are inverse to one another in the class group, we see that direct and inverse composition of forms produce results which are inverse to one another in the class group. Thus, we may again insist on proper equivalence, and we may always choose the $n_i$ positive in composition. The care we have taken up until now in distinguishing the cases has been solely for the purpose of not excluding any kind of composition of forms, to show that the composition we have defined earlier is in a very real sense the only such operation that can exist.

# 7.3  Composition in Different Orders

We now present the general composition algorithm of Arndt.

**Theorem 7.8.** *Let $f_1 = (a_1,\ b_1,\ c_1)$ and $f_2 = (a_2,\ b_2,\ c_2)$ be primitive forms of discriminants $d_1$ and $d_2$, respectively, such that $d_1 = Dn_1^2$ and $d_2 = Dn_2^2$ for integers $n_1$ and $n_2$ and $D$, with $D = \gcd(d_1, d_2)$. Let*

$$m = \gcd\left(a_1 n_2, a_2 n_1, \frac{b_1 n_2 + b_2 n_1}{2}\right).$$

*Then the congruences*

$$mn_1 B \equiv mb_1 \pmod{2a_1}$$

$$mn_2 B \equiv mb_2 \pmod{2a_2}$$

$$m(b_1 n_2 + b_2 n_1)B \equiv m(b_1 b_2 + Dn_1 n_2) \pmod{4a_1 a_2}$$

*are simultaneously solvable for an integer $B$, and the transformation with integer coefficients*

$$\begin{pmatrix} X \\ Y \end{pmatrix}$$

$$= \begin{pmatrix} m & \dfrac{m(b_2 - Bn_2)}{2a_2} & \dfrac{m(b_1 - Bn_1)}{2a_1} & \dfrac{[b_1 b_2 + Dn_1 n_2 - B(b_1 n_2 + b_2 n_1)]m}{4a_1 a_2} \\ 0 & \dfrac{a_1 n_2}{m} & \dfrac{a_2 n_1}{m} & \dfrac{b_1 n_2 + b_2 n_1}{2m} \end{pmatrix} \begin{pmatrix} x_1 x_2 \\ x_1 y_2 \\ y_1 x_2 \\ y_1 y_2 \end{pmatrix}$$

*has determinants*

$$P_{01} = a_1 n_2$$

$$P_{02} = a_2 n_1$$

$$P_{03} = \frac{b_2 n_1 + b_1 n_2}{2}$$

$$P_{12} = \frac{b_2 n_1 - b_1 n_2}{2}$$

$$P_{13} = c_2 n_1$$

$$P_{23} = c_1 n_2$$

*which have no common factor. Under this transformation, the compo-*
*sition of $f_1$ and $f_2$ is*

$$F = \left( \frac{a_1 a_2}{m^2}, \ B, \ \frac{(B^2 - D)m^2}{4a_1 a_2} \right)$$

*of discriminant $D$.*

**Proof.** Given that the congruences are solvable for $B$, the rest is easy. The transformation certainly has integer coefficients and the given discriminants. Any common factor of the determinants must divide $P_{01}$, $P_{23}$, and $P_{03} - P_{12}$ and hence $n_2$. Similarly, any common factor must also divide $P_{02}$, $P_{13}$, and $P_{03} + P_{12}$ and hence $n_1$. Since $n_1$ and $n_2$ are relatively prime by definition, the determinants must therefore have no common factor. The rest is clear from simply performing the calculations.

The proof thus hinges on the solution $B$. If, as in Chapter 4, we let $t$, $u$, $v$ be integers chosen so that

$$a_1 n_2 t + a_2 n_1 u + \left( \frac{b_1 n_2 + b_2 n_1}{2} \right) v = m$$

and set

$$B = \frac{a_1 b_2 t}{m} + \frac{a_2 b_1 u}{m} + \left( \frac{b_1 b_2 + D n_1 n_2}{2m} \right) v,$$

then clearly $B$ is an integer. Further, one can show by direct calculation that $B$ solves the congruences. Specifically,

$$
\begin{aligned}
m n_1 B &= a_1 b_2 n_1 t + a_2 b_1 n_1 u + \left( \frac{b_1 b_2 n_1 + D n_1^2 n_2}{2} \right) v \\
&= a_1 b_2 n_1 t + b_1 \left[ m - a_1 n_2 t - \left( \frac{b_1 n_2 + b_2 n_1}{2} \right) v \right]
\end{aligned}
$$

$$+ \left( \frac{b_1 b_2 n_1 + D n_1^2 n_2}{2} \right) v$$

$$= m b_1 + v \left( \frac{-b_1^2 n_2 + D n_1^2 n_2}{2} \right)$$

$$= m b_1 - 2 a_1 c_1 n_2 v$$

so that the first congruence (and by symmetry the second) is seen to be satisfied. For the third, we have

$$m(b_1 n_2 + b_2 n_1) = (b_1 n_2 + b_2 n_1) \cdot$$

$$\left[ a_1 b_2 t + a_2 b_1 u + \left( \frac{b_1 b_2 + D n_1 n_2}{2} \right) v \right]$$

$$= (b_1 n_2 + b_2 n_1)(a_1 b_2 t + a_2 b_1 u) +$$

$$(b_1 b_2 + D n_1 n_2) \left( \frac{b_1 n_2 + b_2 n_1}{2} \right) v$$

$$= m(b_1 b_2 + D n_1 n_2) + a_1 n_1 t (b_2^2 - D n_2^2) +$$

$$a_2 n_2 u (b_1^2 - D n_1^2)$$

$$= m(b_1 b_2 + D n_1 n_2) + 4 a_1 a_2 (c_2 n_1 t + c_1 n_2 u).$$

The proof is complete. We note that if there were a prime $p$ such that $p \mid a_1$ and $p \mid n_1$, then we would have $p \mid b_1$ and thus that the form would not be primitive. Arguing similarly for the second subscript, we have that

$$m = \gcd \left( a_1 n_2, a_2 n_1, \frac{b_1 n_2 + b_2 n_1}{2} \right) = \gcd \left( a_1, a_2, \frac{b_1 n_2 + b_2 n_1}{2} \right).$$

Of particular special interest in composition is the effect of deriving primitive forms and the relationship between the group of classes of some discriminant $\Delta$ and the group of classes of discriminant $\Delta n^2$. This relationship is explained in the next theorem.

**Theorem 7.9.** *Let $G_\Delta$ and $G_{\Delta n^2}$ be the groups of classes of primitive forms of discriminants $\Delta$ and $\Delta n^2$, respectively. Let $1_\Delta$ be the class of the form $(1,\ 0,\ \Delta/4)$ or $(1,\ 1,\ (1-\Delta)/4)$, whichever is appropriate. Then the map*

$$1_\Delta : G_{\Delta n^2} \to G_\Delta$$

*defined by*

$$1_\Delta : f \to 1_\Delta \circ f$$

*is a group homomorphism.*

**Proof.** We shall prove the theorem in the case of primes $n$. The extension will be obvious. What we need to show is that for any forms $f_1$ and $f_2$ of discriminant $\Delta n^2$, whose composition is $F$, we have

$$(1_\Delta \circ f_1) \circ (1_\Delta \circ f_2) \sim (1_\Delta \circ F).$$

We recall equations (7.5) listing the possible classes of discriminant $\Delta p^2$ for primes $p$

$$(a,\ bp,\ cp^2)$$

$$(ap^2,\ p(b+2ah),\ ah^2 + bh + c).$$

By a single equivalence step we have that

$$(ap^2,\ p(b+2ah),\ ah^2 + bh + c) \sim (ah^2 + bh + c,\ -p(b+2ah),\ ap^2)$$

and can thus assume that any form of discriminant $\Delta p^2$ is equivalent to a form $(a,\ bp,\ cp^2)$ for which $a$ is relatively prime to $p$. Since we know that composition is defined on classes, we can, without loss of generality, assume that $f_1$ and $f_2$ are written in this manner. In the case of even discriminant $\Delta$, when we compound such a form with $1_\Delta$, we have that $m = 1$ and that $B = b$ satisfies the necessary congruences. Thus the composition

$$1_\Delta \circ (a,\ bp,\ cp^2) \sim (a,\ b,\ c).$$

The rest is just a matter of following the algorithm for composition, and the proof for odd discriminants is a simple analogy.

One useful application of the above theorem is in the following corollary.

**Corollary 7.10.** *Let $\Delta$ and $\Delta n^2$ be discriminants of forms. For any prime $q$, the classes of order $q^k$ for all $k$ in the group of discriminant $\Delta n^2$ can all be obtained by deriving either the identity of discriminant $\Delta$ or the classes of order $q^i$ for $i \leq k$ in the group of discriminant $\Delta$.*

**Proof.** Let $F$ be a form of order $q^k$ and discriminant $\Delta n^2$. Then $F^{q^k}$ is the identity $1_{\Delta n^2}$ of discriminant $\Delta n^2$. Thus

$$1_\Delta \circ F^{q^k} \sim (1_\Delta \circ F)^{q^k} \sim 1_\Delta$$

and therefore $1_\Delta \circ F$, the unique class of discriminant $\Delta$ from which $F$ can be derived, is of order some power of $q$ no larger than $k$.

# Chapter 8

# Miscellaneous Facts II

We shall consider in this chapter some basic topics, some of which are clearly interrelated.

First, there are heuristic conjectures by Cohen and Lenstra [COHE84] that, if true, would explain why class numbers and class groups do not appear to be "random" numbers and groups, but have decidedly special properties, among them the strong tendency of the groups to be cyclic.

We shall also present a computational technique for decomposing the finite abelian class groups into their cyclic factors. This technique is successful, in part, because of the observed special nature of class groups.

Also presented will be some results giving conditions under which class numbers possess certain congruence conditions for odd prime moduli, for example, conditions under which there exist elements of order 3 in the group.

In contrast with producing elements of order 2 in class groups, which can be done by choosing highly composite fundamental discriminants, for example, there is no effective way to generate class groups with a large number of elements of a fixed odd order. There has been substantial effort and computation, however, to search for class groups with a large rank in the $p$-Sylow subgroup for odd primes $p$. As is evident from the conjectures of the Cohen-Lenstra heuristics, such a search is less likely to succeed the larger the prime $p$, and most of the effort has gone toward finding class groups of large rank in the 3-Sylow and the 5-Sylow subgroup.

Finally, there is the obvious tantalizing question as to whether there is or should be any connection between the class groups of positive and negative discriminant with similar discriminants. A partial answer has been given by Scholz.

As in the earlier chapter on miscellaneous facts, some of the commentary may involve mathematics which is deeper than the rest of this work. We trust the astute reader can skim the commentary; the statements of the results should be readily comprehensible.

# 8.1 The Cohen-Lenstra Heuristics

We have observed that class groups tend to be cyclic or nearly cyclic. It has also been observed, in part due to interest in a method for factoring integers, that the class numbers $h$ are more often divisible by small prime numbers than are random numbers of comparable size. That is, more than one-third of an arbitrarily chosen set of class numbers will be divisible by 3, and so forth. Henri Cohen and Hendrik Lenstra, Jr., have published a paper with some heuristic explanations and conjectures as to why these observations should hold [COHE84]. In essence, they suggest that the probability of occurrence of specific abelian groups as class groups should be weighted by the number of automorphisms of those groups. Some of their conjectures are presented here.

We consider odd primes $p$ and the odd part $C_0$, of order $h_0$, of the class group of a given discriminant. $\zeta(s)$ is the Riemann $\zeta$-function. $\eta(p)$ is the product $\prod_{i \geq 1}(1 - p^{-i})$. $\lambda$ is the product $\kappa \prod_{i \geq 2} \zeta(i)$, where $\kappa$ is the residue at $s = 1$ of $\zeta(s)$. For the examples, we will need that $\eta(3)$ is about 0.56013, $\eta(5)$ is about 0.76033, and $\eta(7)$ is about 0.83680. Conjectures 1 through 4 refer to negative fundamental discriminants.

*Conjecture 1.*

The probability that $C_0$ is cyclic is

$$\frac{\zeta(2)\zeta(3)}{3\zeta(6)\lambda\eta(2)},$$

which is approximately 0.977575. That is, about 97.7575% of groups $C_0$ are cyclic.

*Conjecture 2.*

The probability that $p$ divides $h_0$ is

$$f(p) = 1 - \eta(p) = \frac{1}{p} + \frac{1}{p^2} - \frac{1}{p^5} - \frac{1}{p^7} + \cdots.$$

Thus

$f(3)$ is approximately 0.43987;

$f(5)$ is approximately 0.23967;

$f(7)$ is approximately 0.16320.

*Conjecture 3.*

If $\mathcal{G}$ is a finite abelian group of odd order, and $q_i$, $i = 1, \ldots, k$ are the primes dividing the order of $\mathcal{G}$, then the probability that the product of the $p_i$-Sylow subgroups of $\mathcal{C}_0$ is isomorphic to $\mathcal{G}$ is

$$\frac{\prod_{i=1}^{k} \eta(p_i)}{\# Aut(\mathcal{G})},$$

where $\# Aut(\mathcal{G})$ is the number of automorphisms of $\mathcal{G}$. For example,

the probability that the 3-Sylow subgroup of $\mathcal{C}_0$

is $C(3)$ is about $0.56013/2 = 0.280065$;

is $C(9)$ is about $0.56013/6 = 0.09335$;

is $C(27)$ is about $0.56013/18 = 0.03112$;

is $C(3) \times C(3)$ is about $0.56013/48 = 0.01167$;

is $C(3) \times C(3) \times C(3)$ is about $0.00005$.

Similarly, the probability that the 5-Sylow subgroup of $\mathcal{C}_0$

is $C(5)$ is about $0.76033/4 = 0.19008$;

is $C(25)$ is about $0.76033/20 = 0.03802$;

is $C(125)$ is about $0.76033/100 = 0.0076033$;

is $C(5) \times C(5)$ is about $0.76033/480 = 0.00158$.

*Conjecture 4.*

The probability that the $p$-Sylow subgroup has rank $r$ is

$$\eta(p)p^{-r^2} \prod_{1 \leq k \leq r} \left(1 - \frac{1}{p^k}\right)^{-2}.$$

We note the combined effects of Conjectures 1 through 4, that the probability that the $p$-Sylow subgroup of a group $\mathcal{C}_0$

has rank 0 is $\eta(p)$;

has rank 1 is $\eta(p) \left(\frac{1}{p}\right) \left(\frac{p}{p-1}\right)^2$;

has rank 2 is $\eta(p) \left(\frac{1}{p^4}\right) \left(\frac{p}{p-1}\right)^2 \left(\frac{p^2}{p^2-1}\right)^2$;

and so forth. The sum of all the multipliers of $\eta(p)$ must be $\frac{1}{\eta(p)}$. Also, it is clear that if these conjectures are in fact true, then the cyclic nature of class groups is explained; the fraction of $p$-Sylow subgroups of rank $r$ for a given prime $p$ is of the order of $\frac{1}{p^{r^2}}$.

For positive fundamental discriminants, we have a different set of conjectures.

*Conjecture 5.*

The probability that $p$ divides $h_0$ is

$$1 - \left(\frac{\eta(p)}{1 - \frac{1}{p}}\right) = \frac{1}{p^2} + \frac{1}{p^3} + \frac{1}{p^4} - \frac{1}{p^7} - \cdots .$$

*Conjecture 6.*

If $\mathcal{G}$ is a finite abelian group of odd order $g$, the probability that $\mathcal{C}_0$ is isomorphic to $\mathcal{G}$ is

$$\frac{1}{2g\lambda\eta(2)\#Aut(\mathcal{G})}.$$

In particular, the probability that $\mathcal{C}_0$

is trivial is 0.75446;

is $\mathcal{C}(3)$ is 0.12574;

is $\mathcal{C}(5)$ is 0.03772;

is $\mathcal{C}(7)$ is 0.01796;

is $\mathcal{C}(9)$ is 0.01572.

*Conjecture 7.*

The probability that the $p$-Sylow subgroup has rank $r$ is

$$\frac{1}{p^{-r(r+1)}} \left( \frac{1}{1 - \frac{1}{p^{r+1}}} \right) \prod_{1 \leq k \leq r} \left( 1 - \frac{1}{p^k} \right)^{-2}.$$

As noted by Cohen and Lenstra, these heuristics fit the statistical data very well (for example, [BUEL84]), and in a few special cases the heuristics have been proved as theorems. Since the publication of the heuristics by Cohen and Lenstra, similar results have appeared (for example, [GERT86, WASH86]).

# 8.2 Decomposing Class Groups

Class groups of forms are finite abelian groups, and we can apply the decomposition theorem for such groups, which we stated in one form in Chapter 4 and which we restate here in a different form.

**Theorem.** *Let $\mathcal{G}$ be a finite abelian group, written multiplicatively, and let $\mathcal{C}(r)$ denote a cyclic group of order $r$. $\mathcal{G}$ can be written as a direct product of groups of prime power order,*

$$\mathcal{G} = \mathcal{S}_{p_1} \times \mathcal{S}_{p_2} \times \cdots \times \mathcal{S}_{p_k},$$

*in which each group $\mathcal{S}_{p_i}$ is of order some power of $p_i$ and is a direct product of cyclic groups each of order some power of $p_i$:*

$$\mathcal{S}_{p_i} = \mathcal{C}(p_i^{\alpha_1}) \times \mathcal{C}(p_i^{\alpha_2}) \times \cdots \times \mathcal{C}(p_i^{\alpha_l}),$$

*where we may choose $\alpha_1 \leq \alpha_2 \leq \ldots \leq \alpha_l$.*

In any group $\mathcal{G}$ of order $h = p^k h'$ with $p$ prime and not dividing $h'$, the set $\mathcal{S} = \{f^{h'} : f \in \mathcal{G}\}$ is a subgroup of order $p^k$ called the *p-Sylow subgroup*. Our decomposition technique for class groups will actually work for any finite abelian group, but is more effective for class groups than for groups in general because of certain specific characteristics of class groups. The approach outlined here is the one used in both computations by Buell [BUEL76] and [BUEL87a], and is derived from the suggestions about computations in class groups made by Shanks [SHAN69] and conveyed to Buell by A. O. L. Atkin.

We decompose the group one $p$-Sylow subgroup at a time. An initial step is to determine whether it is possible for the $p$-Sylow subgroup to be noncyclic. For odd primes $p$, we must have at least $p^2$ dividing $h$.

For the prime 2, we must have at least 4 genera and at least a factor of 16 in the class number (at least a factor of 4 in the number of forms per genus). This is because some basic group structure is inherited from the structure of the genera–a discriminant with $k$ generic characters has exactly $k - 1$ cyclic factors of order some power of 2. It is thus customary, in describing "noncyclic" groups, to consider the 2-Sylow subgroup to be noncyclic if and only if the subgroup of squares (that is, the principal genus) is noncyclic. By squaring, any cyclic factors of order 2 disappear. A "minimally noncyclic" 2-Sylow subgroup is thus a subgroup of the form $\mathcal{C}(4) \times \mathcal{C}(4)$.

Having determined that the class number $h$ is sufficiently composite that some $p$-Sylow subgroup might be noncyclic, we then, for each such prime $p$, choose a few (perhaps a dozen) forms $f$ at random and compute $f^{h/p}$. If any of these is not the identity, the $p$-Sylow subgroup is cyclic. If all are not the identity, there is reason to suspect that the $p$-Sylow subgroup is noncyclic. For the 2-Sylow subgroup, we consider the $fpg$-th power of random forms, where $fpg$ is the number of forms per genus.

Given a class group of order $h = p^k h'$, whose $p$-Sylow subgroup is thought to be noncyclic, we break down the subgroup into its cyclic factors as follows.

*PART I*

1. Choose at random a form $f_1$ of order some power of $p$.

2. Compute the order $p^{ord_1}$ of $f_1$. Save, in a list, the penultimate $p$-power-cycle $f_1^{i(p^{ord_1-1})}$, for $i = 1, \ldots, p - 1$, of $f_1$.

3. Choose at random a form $f_2$ of order some power of $p$.

4. Compute the order $p^{ord_2}$ of $f_2$. Exchange $f_1$ and $f_2$, if necessary, so that $ord_2 \leq ord_1$. If exchange is necessary, compute the penultimate $p$-power cycle of the new $f_1$.

(We now have $f_1$ of order $ord_1$ and $f_2$ of order $ord_2$, with $ord_2 \leq ord_1$.)

5. If $f_2^{p^{ord_2-1}}$ is equal to $f_1^{i(p^{ord_1-1})}$ for any $i$, replace $f_2$ by $f_2 f_1^{-i(p^{ord_1-1})}$ and go to Step 4.

At the end of Part I we have two forms which generate independent cycles. If the sum of $ord_1$ and $ord_2$ is equal to $k$, then we have generated the $p$-Sylow subgroup, and its structure is $C(p^{ord_2}) \times C(p^{ord_1})$. If not, we must compute further.

*PART II*

6. Save, in a list, the penultimate $p$-power-cycle $f_2^{j(p^{ord_2-1})}$, for $j = 1, \ldots, p-1$, of $f_2$, and all cross products $f_1^{i(p^{ord_1-1})} f_2^{j(p^{ord_2-1})}$.

7. Choose at random a form $f_3$ of order some power of $p$.

8. Compute the order $p^{ord_3}$ of $f_3$. If $ord_3 > ord_1$, we replace $f_1$ with $f_3$ and return to Step 3. If $ord_3 > ord_2$, we replace $f_2$ with $f_3$ and return to Step 5.

(We now have forms $f_1$ of order $ord_1$, $f_2$ of order $ord_2$, and $f_3$ of order $ord_3$, with $ord_3 \leq ord_2 \leq ord_1$ and with $f_1$ and $f_2$ generating independent cycles.)

9. If $f_3^{p^{ord_3-1}}$ is equal to $f_1^{i(p^{ord_1-1})} f_2^{j(p^{ord_2-1})}$ for any $i, j$, $0 \leq i, j \leq p - 1$, replace $f_3$ by $f_3 f_1^{-i(p^{ord_1-1})} f_2^{-j(p^{ord_2-1})}$ and go to Step 8.

Some comments are appropriate. In Steps 1, 3, and 7, we choose a form of order some power of $p$ by choosing a form $f$ and computing $f^{h'}$. In theory, we have no guarantee that any given "random" technique for generating forms will not produce elements for which $f^{h'}$ is not always the identity. In practice, we have found that choosing forms whose first coefficients are simply the primes $q$, in order, such that $b^2 \equiv \Delta$ (mod $4q$) is solvable, provides a perfectly adequate list of "random" forms for the purposes of this algorithm.

We also have ignored the obvious exit conditions if we find, in fact, that the group is cyclic. Since the class groups tend to be cyclic about 96% of the time, it is certainly worthwhile to do the test above to determine if a given group is probably noncyclic before beginning the more elaborate decomposition computation. If we ever discover a form $f$ such that $f^{h/p}$ is not the identity, however, then we know that the $p$-Sylow subgroup is cyclic, and we stop.

The heart of this technique is the observation that underlies Steps 5 and 9: If $f_1$ is of order $p^{ord_1}$ and $f_2$ is of order $p^{ord_2}$ with $ord_1 \leq ord_2$, and if $f_2^{p^{ord_2-1}}$ is equal to $f_1^{i(p^{ord_1-1})}$ for some $i$, then $f_2 f_1^{-i(p^{ord_1-1})}$ is of strictly smaller order and is less "dependent" on $f_1$. For example, if we have a group of order 81,

$$\{1, a, a^2\} \times \{1, b, b^2, \dots, b^{26}\}$$

and we choose elements $b$ and $ab^5$, both of order 27, these two elements generate the full group, but not in a way which can be recognized without generating nearly the entire group, a process wasteful both in time and space. By saving the penultimate cycle of $b$, that is, $b^9$ and $b^{18}$, we can determine that the penultimate $p$-power of $ab^5$, which is $(ab^5)^9 = b^{18}$, matches one of our saved elements, and $(ab^5)b^{-2} = ab^3$

is of order 9 and not 27. In the second pass through Step 5 we will replace $ab^3$ by $a$ and have independent generators.

If we fall through Part II entirely, we have three forms which generate three independent cycles. This is, as we have seen statistically, very uncommon for class groups. Almost always it will be the case that we will have $ord_1 + ord_2 + ord_3 = k$, and thus that we have decomposed the entire $p$-Sylow subgroup. If not, we have several options for continuing the decomposition. One approach would be to save the penultimate $p$-power cycle for the new form $f_3$ and for all the cross terms, generate a fourth form $f_4$, and try to reduce it to an independent fourth generator. This is, however, very unlikely; it is much more likely that in the process of reduction we will reduce the new form to a generator of a cycle which subsumes one of the three already known to exist.

A second approach, which requires less complicated programming, would be to abandon completely the idea of reducing forms to generators of cycles, to assume the existence of three (and not four or more) cyclic factors, and try to build the subgroup on this basis. For example, if $3^5$ exactly divides $h$ and we have generated a subgroup $\mathcal{C}(3) \times \mathcal{C}(3) \times \mathcal{C}(9)$, then there are only two possibilities–the subgroup is either $\mathcal{C}(3) \times \mathcal{C}(9) \times \mathcal{C}(9)$ or $\mathcal{C}(3) \times \mathcal{C}(3) \times \mathcal{C}(27)$. The difference between the two is in the number of elements of order 9. In the former group, we have 80 such, while in the latter we have only 8. Computing the class groups of negative discriminant for discriminants from $-4$ to $-25000000$ yielded only about 45 instances (the exact number depended on versions of the program) in which the subgroup was determined to have three or more cyclic factors but was not completely determined by the computations of Parts I and II.

A third approach, of course, would be to completely abandon the existing computation and start completely from the beginning, with a new list of random forms, in hopes that this time the full group would be determined in Parts I and II.

Class groups, as we have seen, tend to be cyclic, and if not cyclic, not to be very noncyclic. That is, among noncyclic $p$-Sylow subgroups of order $p^4$, a class group $\mathcal{C}(p) \times \mathcal{C}(p^3)$ is far more likely to be encountered than a group $\mathcal{C}(p^2) \times \mathcal{C}(p^2)$. One reason for the success of this decomposition method is that it clearly works best on groups of this nature, for which it is statistically likely to be able to generate most, or all, of the group with exactly two elements, provided one element is reduced to eliminate dependence on the other.

# 8.3  Specifying Subgroups of Class Groups

Some obvious questions raised about class numbers and class groups are these: Are there simple conditions under which one can guarantee that the class number of a certain discriminant is divisible by a given odd prime $p$? Are there effective conditions for this, in the sense that one can obtain a computational answer for a given discriminant in significantly less time than it would take to generate the entire group? Can one classify discriminants whose class numbers are divisible by $p$? Are there infinitely many discriminants whose class numbers are divisible (or not divisible) by $p$? The answers to these questions, with some qualifications (more for positive discriminants than for negative discriminants, as might be imagined), is yes.

## 8.3.1  Congruence Conditions

Most results come from the following approach. If, for example, for a discriminant $\Delta$, we have

$$b^2 - 4a^3 = \Delta, \quad \text{with} \quad \gcd(a, b) = 1 \tag{8.1}$$

then clearly we have a primitive form $f = (a, b, a^2)$ whose square is $(a^2, b, a) \sim (a, b, a^2)^{-1}$. Thus $f$ is either principal or of order 3. If one can guarantee that it is not principal, we have explicitly given a form of order 3 in the class group. For negative discriminants, for example, this is merely a condition on the magnitude of $a$. For positive discriminants, there is the ever-present problem of having many reduced forms in a given cycle and class. It is also the case, however, that one must consider not just fundamental but also derived discriminants, remembering from Corollary 7.10 that forms of prime order $p$

of derived discriminant $\Delta n^2$ must be derived either from forms of order $p$ or from the principal form of discriminant $\Delta$. Nagell proved in 1922 that there are infinitely many negative fundamental discriminants whose class numbers are divisible by a given integer $n$ [NAGE22]. We take Theorems 8.1 through 8.4 from a paper by Yamamoto [YAMA70].

**Theorem 8.1.** *Let $n$ be a natural number and $\Delta$ a fundamental discriminant not equal to $-3$ or $-4$. If the triple $[X, Y, Z] = [a, b, e]$ is a solution to the equation*

$$Y^2 - 4X^n = Z^2 \Delta \qquad\qquad (8.2)$$

*for which $\gcd(a, b) = 1$, then the primitive form $F = (a, \ b, \ a^{n-1})$*

*of discriminant $\Delta e^2$ corresponds to a primitive form of discriminant $\Delta$ whose $n$-th power is principal and which is not equivalent to its inverse.*

From this we infer that all the elements of order $n$ in the class group must be obtainable from the solutions of (8.2). Not all solutions lead to nonprincipal classes, and not all solutions lead to inequivalent classes, but solving (8.2) is necessary.

To describe the theorems, we now need to extend the notion of congruence and the notion of residuacity. In the rational integers, we say that $a$ is congruent to $b$ modulo $m$ if $m$ divides $a - b$. In terms of ideals, to divide is to contain, and we can define congruence modulo ideals in the rational integers by saying that $a$ is congruent to $b$ modulo $m$ if $a - b$ is contained in the ideal $\mathbf{m} = \{km : k \in \mathbf{Z}\}$ in the rational integers. Similarly, for integers in a fixed quadratic (or any algebraic number) field, we can say that $\alpha$ is congruent to $\beta$ modulo the ideal **m** if $(\alpha - \beta) \in$ **m**.

Power residues can be described similarly. In the rational integers, $a$ is an $n$-th power residue modulo $m$ if there exists an integer $b$ such that $b^n \equiv a \pmod{m}$. In terms of ideals in a fixed quadratic number field, we can say that a quadratic algebraic integer $\alpha$ is an $n$-th power residue modulo an ideal **m** if there exists a quadratic algebraic integer $\beta$ such that $(\beta^n - \alpha) \in$ **m**.

We recall from the last part of Chapter 6 that the form $(a, b, c)$ of fundamental discriminant $\Delta$ corresponds to the ideal $< a, \frac{-b+\sqrt{\Delta}}{2} >$ of the quadratic field $\mathbf{Q}(\sqrt{\Delta})$, and that this correspondence is an isomorphism of the classes of forms and narrow ideals. In what follows we also let $\varepsilon$ denote the fundamental unit of the quadratic field $\mathbf{Q}(\sqrt{\Delta})$ if $\Delta$ is positive or simply 1 if $\Delta$ is negative. With these definitions and notation we have the following theorem.

**Theorem 8.2.** *Let* $n = p_1^{u_1} p_2^{u_2} \ldots p_s^{u_s}$ *be the canonical decomposition of* $n$ *into prime power factors. Choose for each* $i$ *a prime* $q_i$ *such that*

$q_i \equiv 1 \pmod{p_i}$ *for odd primes* $p_i$,

$q_i \equiv 1 \pmod{4}$ *if* $p_i = 2$.

*Let* $[a, b, e]$ *is a solution of (8.2) as above, with* $\gcd(a, b) = 1$. *If*

*i)* $q_i$ *divides* $a$ *for* $i = 1, \ldots, s$;

*ii)* $b$ *is a* $p_i$-*th power nonresidue modulo* $q_i$, *for* $i = 1, \ldots, s$;

*iii)* $\varepsilon$ *is a* $p_i$-*th power residue modulo* $\mathbf{q}_i$, *for* $i = 1, \ldots, s$, *where* $\mathbf{q}_i$ *is one of the two ideals corresponding to the forms* $(q_i, *, *)$ *of discriminant* $\Delta$, *then the form* $F$ *obtained from the solution to (8.2) is not principal and is of order* $n$ *in the class group.*

*Remark:* In the case of negative discriminant, condition iii) is always met.

From this theorem it is a short step to the next.

**Theorem 8.3.** *Let $S_1$, $S_2$, $S_3$, be sets of rational prime numbers such that $S_i \cap S_j = \emptyset$ . There exist infinitely many negative fundamental discriminants of forms $\Delta$ for which*

  a) *the class group of forms has a class of order n;*

  b) *for each prime p in the set $S_1$ there are pairs of inequivalent classes of forms $(p, k, l)$, $(p, -k, l)$ of lead coefficient p and discriminant $\Delta$;*

  c) *for no prime p in the set $S_2$ does there exist a form of lead coefficient p and discriminant $\Delta$;*

  d) *each prime p in the set $S_3$ divides $\Delta$.*


We shall not completely prove Theorem 8.3, but will give some insight as to the computations involved. For simplicity, we exclude 2 and 3 from $S_3$. We let $S$ denote the set of prime factors of $n$. We choose a fixed prime $q$ such that

  $q \equiv 1 \pmod{q}$ for $q \in S \cup S_2$;

  $q \equiv 1 \pmod{q^2}$ for $q \in S_3 \cup \{2\}$.

Let $k$ be the product of the primes in $S$. We choose a rational integer $b$ such that

  i) $b$ is a $p_i$-th power nonresidue modulo $q$ for all $p_i \in S$;

  ii) $\gcd(b, k) = 1$;

iii) $b^2 - 4$ is a quadratic nonresidue modulo the odd primes $p \in S_2$ and is odd if $2 \in S_2$;

iv) $b \equiv p + 2 \pmod{q^2}$ for $q \in S_3$.

We choose $a$ such that

v) $\gcd(a, b) = 1$;

vi) $kq \mid a$;

vii) $a \equiv 1 \pmod{q^2}$ for $q \in S_2 \cup S_3$;

viii) $b^2 - 4a^n < 0$.

Such integers $a$ and $b$ can be easily guaranteed to exist, since we have linear congruence conditions and can cite Dirichlet's theorem on primes in progressions if necessary. We then have nonfundamental discriminants $\Delta e^2 = b^2 - 4a^n$ from which we can extract an infinite list of fundamental discriminants with the desired properties.

Yamamoto also shows in his paper how two triples of solutions to (8.2) can be guaranteed to exist, by extensions of these congruence conditions, and generate independent cycles of order 1 or $n$. Since only one such could be the identity in the case of positive discriminant, he proves the following theorem.

**Theorem 8.4.** *Let $S_1$, $S_2$, $S_3$, be sets of rational prime numbers such that $S_i \cap S_j = \emptyset$ . There exist infinitely many negative (respectively, positive) fundamental discriminants of forms $\Delta$ for which*

a) *the class group of forms has a subgroup isomorphic to $C(n) \times C(n)$; (respectively $C(n)$ );*

b) *for each prime p in the set $S_1$ there are pairs of inequivalent classes of forms $(p, k, l)$, $(p, -k, l)$ of lead coefficient p and discriminant $\Delta$;*

c) *for no prime p in the set $S_2$ does there exist a form of lead coefficient p and discriminant $\Delta$;*

d) *each prime p in the set $S_3$ divides $\Delta$.*

We thus have shown that infinitely many discriminants exist with class numbers divisible by $n$ for any given integer $n$. In addition, we have shown that infinitely many $C(n) \times C(n)$ subgroups occur for negative discriminants, and infinitely many $C(n)$ subgroups occur for positive discriminants. A common phenomenon appears here, which will also appear later in this chapter. The technique produces explicit forms of order $n$ or 1, and in the case of Theorem 8.4, produces two such forms from independent cycles. From this we can conclude that we have two cyclic factors of order $n$ in the case of negative discriminant. For positive discriminant, one of these forms–but only one–could be principal, and we have guaranteed the existence of exactly one fewer cyclic factor than in the case of positive discriminant. This will appear later in discussing Scholz's result on the 3-Sylow subgroup and Shanks' use of Scholz's result, and should be compared with the Cohen-Lenstra heuristics. The difference in the heuristics for positive and negative discriminants is essentially a factor of $\frac{1}{p}$, and is the direct result of the existence of the nontrivial units in the ring of integers.

The technique used by Yamamoto is standard for proving results guaranteeing certain forms to exist in class groups: we use for the discriminant a polynomial which has certain explicit algebraic factorings or properties. Weinberger has done this in an alternate proof of the existence of infinitely many positive fundamental discriminants with class number divisible by a given integer $n$ [WEIN73]. Weinberger shows that if we let the discriminant $\Delta$ be the fundamental part of $\Delta(x) = x^{2n} + 4$, then the class number of $\Delta$ is infinitely often divisible by $n$. Honda earlier showed [HOND68] that if we let $\Delta$ be the fundamental part of discriminants $4x^3 - 27y^2$ for an appropriate choice of $x$ and $y$, then for infinitely many positive $\Delta$ we have class numbers divisible by 3. Similarly, Chowla and Hartung [CHOW74] showed that for discriminants of the form $\Delta = -(27n^2 + 4)$ for which $-\Delta$ is prime, the class number is divisible by 3. The opposite question has also been studied by Hartung, who showed [HART74] that there exist infinitely many negative fundamental discriminants whose class numbers are *not* divisible by 3.

Unfortunately, as a moment's thought will show, (8.2) is not entirely practical as a means of generating classes of order $n$ in class groups. The problem is twofold. First, the equation itself is open-ended, as there are no conditions of magnitude on any of the variables. Second, the equation is of degree $n$, so that the computational question of handling the arithmetic in searching for solutions is significant. For these reasons, solving (8.2) has usually been used as a heuristic in finding forms of some order $n$ rather than a method for exhaustively enumerating them.

## 8.3.2 Exact and Exotic Groups

Having guaranteed by various means that noncyclic $p$-Sylow subgroups exist for all $p$, at least for negative discriminants, it is not necessarily trivial to find examples even for all small primes $p$. In the following table we list for the small primes $p$ the first occurrence, for even and odd negative discriminants, of a noncyclic $p$-Sylow subgroup (the notation in the third column $3 \times 9$, for example, indicates a group $\mathcal{C}(3) \times \mathcal{C}(9)$) [BUEL76, BUEL87a].

| Prime | Discriminant | Complete Class Group |
|-------|-------------|---------------------|
| 3 | −3299 | 3 × 9 |
| | −3896 | 3 × 12 |
| 5 | −11199 | 5 × 20 |
| | −17944 | 5 × 10 |
| 7 | −63499 | 7 × 7 |
| | −159592 | 7 × 14 |
| 11 | −65591 | 11 × 22 |
| | −580424 | 22 × 22 |
| 13 | −228679 | 13 × 26 |
| | −703636 | 13 × 26 |
| 17 | −1997799 | 34 × 34 |
| | −4034356 | 17 × 34 |
| 19 | −373391 | 19 × 38 |
| | −3419828 | 19 × 38 |
| 23 | −7472983 | 23 × 46 |
| | −11137012 | 23 × 46 |
| 29 | −20113607 | 29 × 116 |
| | −16706324 | 58 × 58 |
| 31 | −11597903 | 31 × 62 |
| 41 | −6112511 | 41 × 82 |

The discussion of this section has centered on subgroups of the class group; there is also some small interest in knowing which finite abelian groups actually occur as the exact class group, not just a subgroup. Chowla proved that not all elementary 2-groups can occur as class groups of quadratic fields [CHOW34]. In our computation of class groups of negative discriminant, we found that all abelian groups of rank two and order less than 1000 occurred as class groups except $\mathcal{C}(11) \times \mathcal{C}(11)$, $\mathcal{C}(19) \times \mathcal{C}(19)$, $\mathcal{C}(29) \times \mathcal{C}(29)$, and $\mathcal{C}(31) \times \mathcal{C}(31)$ [BUEL87a]. It is therefore almost certain that $\mathcal{C}(11) \times \mathcal{C}(11)$ occurs as the exact class group for *no* negative discriminant. A discriminant $\Delta$ with $-\Delta > 25000000$ and class number as small as 121 would give a value for $L(1,\chi)$ so small (see Chapter 5) as to lie far outside the Littlewood bounds. This is of course permissible–the Littlewood bounds are asymptotic bounds–but would run counter to almost all other experience with class groups.

We know that infinitely many discriminants have $\mathcal{C}(p)$ subgroups of their class groups for all odd primes $p$, and that infinitely many negative discriminants have $\mathcal{C}(p) \times \mathcal{C}(p)$ subgroups. Generating such subgroups is difficult, however. What is even more difficult is to find discriminants whose class groups contain subgroups of rank three or more $\mathcal{C}(p) \times \mathcal{C}(p) \times \mathcal{C}(p)$. Prior to 1970 or so it had even been suggested that such subgroups did not exist. In the early 1970's, however, Maurice Craig showed the existence of a negative fundamental discriminant whose class group had a subgroup $\mathcal{C}(3) \times \mathcal{C}(3) \times \mathcal{C}(3)$ [CRAI73]. In the next few years Shanks, with and without collaborators, Buell, Solderitsch, and primarily Diaz y Diaz found large numbers of discriminants of 3-rank three or four [BUEL76, DIAZ74, DIAZ78, DIAZ79, NEIL74, SHAN72a, SHAN72b, SHAN73b, SOLD77].

The basic technique utilizes Theorem 8.1 in a special form.

**Theorem 8.5.** *If $\Delta$ is a negative fundamental discriminant, then the classes of order 3 in the class group correspond precisely to the solutions of*

$$b^2 - 4a^3 = e^2\Delta$$

*for which $a < \sqrt{\Delta/3}$ and the greatest common divisor $g = \gcd(a, e)$ divides $\Delta$ and is squarefree.*

We use this to find discriminants $\Delta e^2$ which can be represented as

$b^2 - 4a^3$ in a number of different ways, usually by finding polynomials $b_1(x)$, $b_2(x)$, $a_1(x)$, $a_2(x)$, such that

$$\Delta(x) = b_1(x)^2 - 4a_1(x)^3 = b_2(x)^2 - 4a_2(x)^3$$

With appropriate conditions on the sizes of terms, one can guarantee in this way that for negative discriminants we have at least rank two in the 3-Sylow subgroup. Shanks generally took the approach of finding a few very select polynomials, often choosing them so as to be able to apply Scholz's theorem, mentioned at the end of this section. Buell and Diaz y Diaz by contrast used a large number of polynomials. Diaz y Diaz, who was most successful, generated at one point hundreds of discriminants of 3-rank at least two from a large number of polynomials, and then sorted the discriminants in order to find matches. If a discriminant had been generated by different polynomials, with different 3-cycles, it would have rank three or four. More recently Schoof [SCHO83] has found groups with large 3-rank and 5-rank by using the theory of elliptic curves.

We summarize the first occurrences of large rank. The first four lines represent discriminants known to be the first occurrence of 3-rank or of 5-rank three, as all previous class groups have been computed. The remaining examples are merely the first known occurrences.

| Discriminant | Complete Class Group |
|---|---|
| $-3321607$ | $3 \times 3 \times 63$ |
| $-4447704$ | $6 \times 6 \times 24$ |
| $-18397407$ | $5 \times 10 \times 40$ |
| $-11203620$ | $10 \times 10 \times 10$ |
| $-653329427$ | $3 \times 3 \times 3 \times 210$ |
| $-2520963512$ | $3 \times 3 \times 6 \times 276$ |
| $-258559351511807$ | $5 \times 5 \times 10 \times 59140$ |

In computing 3-Sylow subgroups, a significant theorem of Arnold Scholz has been extensively used, especially by Shanks. The theorem relates the class groups of the quadratic fields of radicands $+\Delta$ and $-3\Delta$ [SCHO32].

**Theorem 8.6.** *Let $+\Delta$ and $-3\Delta$ be the radicands of the quadratic fields $\mathbf{Q}\sqrt{\Delta}$ and $\mathbf{Q}\sqrt{-3\Delta}$, where $\Delta$ may or may not be divisible by 3. If $r$ is the rank of the 3-Sylow subgroup of the field with negative discriminant, and $s$ is the rank of the 3-Sylow subgroup of the field with positive discriminant, then*

$$s \leq r \leq s+1.$$

# Chapter 9

# The 2-Sylow Subgroup

One of the most carefully studied questions about quadratic class groups is the precise determination of the 2-Sylow subgroup. This is intimately connected, in the case of positive discriminant, with the question of which discriminants $\Delta$ possess solutions of the negative Pell equation

$$X^2 - \Delta Y^2 = -4 \tag{9.1}$$

and both questions are related for all discriminants to the existence of higher-order reciprocity laws analogous to the law of quadratic reciprocity. The connections come from the following observations. Given a fundamental discriminant with $t$ prime factors, we are guaranteed a factor of $2^{t-1}$ in the class number. This is the exact power of 2 in the class number if and only if the 2-Sylow subgroup is an elementary 2-group, or if the genera each have an odd number of classes, or if there is exactly one ambiguous class per genus, these being equivalent conditions. The ambiguous classes correspond to factors of the discriminant, and a class which represents a prime $p$ dividing $\Delta$ also represents $\Delta/p$, so the 2-Sylow subgroup can only be elementary if no prime factor of $\Delta$ is a quadratic residue of all the other prime factors of $\Delta$ (For the sake of argument, we ignore for the moment the extra characters $\chi$ and $\psi$.) Conversely, if by residue conditions one can guarantee that more than one ambiguous class lies in the principal genus, then that genus and all other genera have an even number of classes, there exist classes of order 4 in the group (since the ambiguous class of order 2 in the principal genus must itself be a square) and thus both a more complex group structure and a higher power of 2 in the class number are assured. Finally, it has been known since the time of Dirichlet that the above matters are related to fourth-power and higher-power residue conditions on the factors of $\Delta$. These have in recent years been the basis for higher-order reciprocity laws.

We shall in general concentrate on positive discriminants and Pell's equation. However, many of the results mentioned here are the links among the arithmetics of the quadratic fields $\mathbf{Q}(\sqrt{m})$, $\mathbf{Q}(\sqrt{n})$, $\mathbf{Q}(\sqrt{mn})$, and the biquadratic field $\mathbf{Q}(\sqrt{m}, \sqrt{n})$ and thus imply results for negative discriminants as well.

As with the previous chapter, we shall not rigorously prove many of the results mentioned. In some instances, the proofs are deep and beyond the scope of this work. In many instances even the statements of the theorems are tedious to detail. Further, it is unfortunately the case that there is no all-inclusive general result. We have theorems covering special cases, and theorems covering all except the most difficult cases, but no unified answer. Our treatment, then, must be somewhat less than perfect; we have tried to present characteristic and understandable results and provide a complete list of references for the literature. Since this list is extensive, we have made it a special list at the end of this chapter. Papers specifically cited by name in this chapter are included for completeness both in the special and the general lists.

Also, similar to the previous chapter, parts of this chapter are at a level higher than other parts of this book, and we will occasionally mention techniques and theory of deeper mathematics than is used elsewhere. The careful reader should be able to skim the commentary and concentrate on understanding the statements of the theorems.

We begin with a general theorem of Gauss.

**Theorem 9.1.** *Let $\Delta$ be a fundamental discriminant. The ambiguous classes are themselves all squares, and thus in the principal genus, if and only if $\left(\frac{p}{q}\right) = +1$ for all pairs $p, q$ of odd primes dividing $\Delta$, and $\left(\frac{2}{p}\right) = +1$ for all odd primes $p$ dividing $\Delta$ if $\Delta$ is even. If this is true, then the 4-rank of the class group is exactly $k-1$, where $k$ is the number of primes dividing $\Delta$.*

As examples, we note that $\Delta = -550712 = -8 \cdot 23 \cdot 41 \cdot 73$ has class group $\mathcal{C}(4) \times \mathcal{C}(4) \times \mathcal{C}(16)$ and $\Delta = -568888 = -8 \cdot 17 \cdot 47 \cdot 89$ has class group $\mathcal{C}(4) \times \mathcal{C}(4) \times \mathcal{C}(8)$.

# 9.1   Classical Results on the Pell Equation

**Theorem 9.2.** *For positive fundamental discriminants* $\Delta$, *the negative Pell equation (9.1) is solvable only if all odd prime divisors of* $\Delta$ *are congruent to* 1 *modulo* 4.

**Proof.** For any odd prime $p$ dividing $\Delta$, (9.1) becomes the congruence $X^2 \equiv -4 \pmod{p}$ so that $-4$ and hence $-1$ must be quadratic residues of $p$, implying that $p$ is congruent to 1 modulo 4.

A better proof, which gets to the heart of the Pell equation problem, is this. The negative Pell equation is solvable if and only if the reduced form with lead coefficient $-1$ is the *other* ambiguous form in the principal cycle. This can only happen if first of all the form $(-1, k_0, *)$ lies in the principal genus. Since the principal genus is the genus for which all the generic characters (which are the quadratic characters modulo the primes $p_i$ dividing $\Delta$ ) are $+1$, we must have that $-1$ is a quadratic residue of all these primes $p_i$.

**Theorem 9.3.** *If* $\Delta = p$, *with* $p$ *a prime congruent to* 1 *modulo* 4, *then equation (9.1) is solvable, and the class number is odd.*

**Proof.** There are, for prime discriminant, exactly two reduced ambiguous forms, those with lead coefficients $+1$ and $-1$. These must therefore lie in the same cycle, so that (9.1) is solvable, and since there are no classes of order 2, the class number must be odd.

**Theorem 9.4.** *If* $\Delta = 8p$, *with* $p$ *a prime congruent to* 5 *modulo* 8, *then equation (9.1) is solvable, and the class number is twice an odd number.*

**Proof.** In this case we have four ambiguous forms distributed over the genera as follows.

$$
\begin{array}{cc}
\left(\tfrac{r}{p}\right) & \psi \\
+ & + \quad (1, *, *), (-1, *, *) \\
- & - \quad (2, *, *), (-2, *, *)
\end{array}
$$

We find again that $(-1, *, *)$ must be in the principal cycle, and since no ambiguous class except the principal class is a square, the class number is twice an odd number.

**Theorem 9.5.** *If $\Delta = pq$, with $p$ and $q$ primes such that $p \equiv q \equiv 1$ (mod 4) and $\left(\tfrac{p}{q}\right) = \left(\tfrac{q}{p}\right) = -1$, then equation (9.1) is solvable, and the class number is twice an odd number.*

**Proof.** Assuming that we have $p < q$, then we have reduced ambiguous forms $(\pm 1, k_0, *)$ and $(\pm p, pk_p, *)$, with the latter pair equivalent to the forms $(\pm q, qk_q, *)$. The reduced ambiguous forms must distribute over the genera as

$$
\begin{array}{cc}
\left(\tfrac{r}{p}\right) & \left(\tfrac{r}{q}\right) \\
+ & + \quad (1, *, *), (-1, *, *) \\
- & - \quad (p, *, *), (-p, *, *)
\end{array}
$$

and thus, as before, the $-1$ form is in the principal cycle and the class number is twice an odd number.

Theorems 9.3, 9.4, and 9.5 are the first examples of a general theorem.

**Theorem 9.6.** *If $\Delta$ is odd and all its prime factors are congruent to 1 modulo 4, or if $\Delta$ is 8 times an odd number all of whose prime factors are congruent to 1 modulo 4, and if the 2-Sylow subgroup is an elementary 2-group, then (9.1) is solvable.*

**Corollary 9.7.** *If* $\Delta = pqr$, *with* $p$, $q$, *and* $r$ *all prime,* $p \equiv q \equiv r \equiv 1$ (mod 4), *and if the three residue symbols* $\left(\frac{p}{q}\right)$, $\left(\frac{p}{r}\right)$, $\left(\frac{q}{r}\right)$, *are either* $(+1, -1, -1)$ *or* $(-1, -1, -1)$, *then (9.1) is solvable.*

**Corollary 9.8.** *If* $\Delta = 8pq$, *with* $p$ *and* $q$ *prime,* $p \equiv q \equiv 1$ (mod 4), *and either*

*a)* $p \equiv q + 4$ (mod 8) *and* $\left(\frac{p}{q}\right) = -1$ *or*

*b)* $p \equiv q \equiv 5$ (mod 8),

*then (9.1) is solvable.*

The theorem and its corollaries are easily seen to be true. Unfortunately, the theorem is more than sufficient, since its conditions guarantee that no reduced ambiguous forms are squares except those of lead coefficients $+1$ or $-1$.

The first open case is now that of discriminants $\Delta = 8p$ for primes $p$ congruent to 1 modulo 8. In this case the four reduced ambiguous forms $(\pm 1, *, *), (\pm 2, *, *)$, all lie in the principal genus. Exactly two are in the principal cycle, so that exactly one of equations (9.1) and

$$2x^2 - py^2 \;=\; +1 \qquad\qquad (9.2)$$
$$-2x^2 + py^2 \;=\; +1 \qquad\qquad (9.3)$$

is solvable. Theorems 9.9 and 9.10 are due to Dirichlet [DIRI]

**Theorem 9.9.**

*a) If (9.2) is solvable, then (9.1) is not solvable, and $p$ is congruent to 1 modulo 16.*

b) *If (9.3) is solvable, then (9.1) is not solvable, and p is a prime for which 2 is a fourth power modulo p.*

c) *If p is a prime congruent to 9 modulo 16 for which 2 is not a fourth power, then (9.1) is solvable.*

**Proof.** If (9.2) is solvable, then we must have $\left(\frac{2}{q}\right) = +1$ for each prime factor $q$ of $y$. Thus all prime factors of $y$ are either $+1$ or $-1$ modulo 8, so that $y^2$ is 1 modulo 16. This implies that $x$ is odd, and thus that $2x^2$ is congruent to 2 modulo 16. Inserting this information in (9.2), we have that $p$ must be 1 modulo 16, proving part a.

If (9.3) is solvable, then $x$ must be even. If we write $x = 2^\alpha q_1 \dots q_k$ for odd primes $q_i$, we have (remembering reciprocity) that $\left(\frac{2^\alpha}{p}\right) = \left(\frac{q_i}{p}\right) = +1$ for all $i$, and thus that $\left(\frac{x}{p}\right) = +1$. We now exponentiate (9.2).

$$(2x^2 - py^2)^{\frac{p-1}{4}} \equiv 2^{\frac{p-1}{4}} x^{\frac{p-1}{2}} \equiv 2^{\frac{p-1}{4}} \equiv 1 \pmod{p}.$$

Now, $2^{\frac{p-1}{4}} \equiv 1 \pmod p$ for primes congruent to 1 modulo 16. We have proved part b. Part c follows from excluding the conclusions of a and b.

As Dirichlet notes, neither condition of part c is necessary. For $\Delta = 8 \cdot 113$ we have $30^2 - 8 \cdot 113 \cdot 1^2 = -4$, but $113 = 16 \cdot 7 + 1$ and $2^{\frac{113-1}{4}} \equiv 1 \pmod{113}$.

It is convenient at this point to introduce the notation for higher residue symbols. If $a$ and $b$ are integers such that $a$ is a quadratic residue of $b$, we can define the symbol $\left(\frac{a}{b}\right)_4$ equal to $+1$ or $-1$ according as $a$ is or is not a fourth power modulo $b$. (One can in general define the symbols as roots of unity, thus being able to remove the initial condition that $a$ is a quadratic residue of $b$, but that is unnecessary for our purposes. We shall only need symbols which can be defined rationally to be either $+1$ or $-1$.)

The final "elementary" case we shall consider in detail is that of discriminant $\Delta = pq$, for $p$ and $q$ prime, both congruent to 1 modulo 4, and for which $\left(\frac{p}{q}\right) = \left(\frac{q}{p}\right) = +1$. Without loss of generality we assume that $p$ is smaller than $q$. As was the case in Theorem 9.9, the reduced ambiguous forms $(\pm 1, *, *)$, $(\pm p, *, *)$, are all in the principal genus, and exactly one of equations (9.1) and the following is solvable.

$$x^2 - pqy^2 = +4p \tag{9.4}$$

$$x^2 - pqy^2 = -4p \tag{9.5}$$

We consider only (9.4), noting that we must have $x = px'$, so that the equation becomes

$$px'^2 - qy^2 = +4$$

There are two possible cases. We may have $x'$ and $y$ both even, so that we have

$$px^2 - qy^2 = +1,$$

or we may have $x'$ and $y$ both odd, with $p$ and $q$ not congruent to each other modulo 8.

In the former case, we first see that $x$ is odd and $y$ even, and as in Theorem 9.9 that $\left(\frac{r}{p}\right) = +1$ for all odd primes $r$ dividing $y$. If $2^\alpha$ is the exact power of 2 dividing $y$, then if $p$ is congruent to 5 modulo 8, we have that $\alpha$ must be 1 and that $\left(\frac{2^\alpha}{p}\right) = \left(\frac{2}{p}\right) = -1$. If $p$ is 1 modulo 8 then $\left(\frac{2}{p}\right) = +1$. Either way, we have $\left(\frac{2^\alpha}{p}\right) = (-1)^{\frac{p-1}{4}}$, so that $\left(\frac{y}{p}\right) = (-1)^{\frac{p-1}{4}}$. Exponentiating (9.4) as in Theorem 9.9, we find that

$$q^{\frac{p-1}{4}} \equiv 1 \pmod{p} \tag{9.6}$$

In the latter case of $x'$ and $y$ both odd we again conclude that (9.6) holds, with only a slight change in the argument. We have proved the following theorem.

**Theorem 9.10.** *If* $\Delta = pq$, *with* $p$ *and* $q$ *primes such that* $p \equiv q \equiv 1$ (mod 4) *and* $\left(\frac{p}{q}\right) = \left(\frac{q}{p}\right) = +1$, *and if* $\left(\frac{p}{q}\right)_4 = \left(\frac{q}{p}\right)_4 = -1$, *then equation* *(9.1) is solvable.*

Again we note, using as example $p = 5$ and $q = 521$, that these conditions are not necessary.

We can summarize these "elementary" results.

**Odd discriminants for which the negative Pell equation is solvable:**

All primes $p$, $q$, $r$ are congruent to 1 modulo 4.

1. $p$: (Theorem 9.3) $h$ is odd;

2. $pq$: $\left(\frac{p}{q}\right) = -1$ (Theorem 9.5) $2 \parallel h$;

3. $pq$: $\left(\frac{p}{q}\right) = +1$, $\left(\frac{p}{q}\right)_4 = \left(\frac{q}{p}\right)_4 = -1$ (Theorem 9.10);

4. $pqr$: $\left(\frac{p}{q}\right) = \left(\frac{p}{r}\right) = \left(\frac{q}{r}\right) = -1$ (Corollary 9.7);

5. $pqr$: $\left(\frac{p}{q}\right) = +1$, $\left(\frac{p}{r}\right) = \left(\frac{q}{r}\right) = -1$ (Corollary 9.7);

6. $pqr$: $\left(\frac{p}{q}\right) = \left(\frac{p}{r}\right) = +1$, $\left(\frac{q}{r}\right) = -1$ (Dirichlet);

7. $pqr$: $\left(\frac{p}{q}\right) = \left(\frac{p}{r}\right) = \left(\frac{q}{r}\right) = +1$;

$\left(\frac{pq}{r}\right)_4 = \left(\frac{pr}{q}\right)_4 = \left(\frac{qr}{p}\right)_4 = \left(\frac{p}{q}\right)\left(\frac{p}{r}\right)_4 = \left(\frac{q}{p}\right)\left(\frac{q}{r}\right)_4 = \left(\frac{r}{p}\right)\left(\frac{r}{q}\right)_4 = -1$ (Dirichlet).

The last two entries are proved in the reference by Dirichlet cited above
by methods similar to Theorems 9.9 and 9.10.

**Even discriminants for which the negative Pell equation is
solvable:**

All primes $p$, $q$, $r$ are congruent to 1 modulo 4.

8. $8p$: $p \equiv 5$   (mod 8) (Theorem 9.4);

9. $8p$: $p \equiv 9$   (mod 16), $\left(\frac{2}{p}\right)_4 = -1$ (Theorem 9.9);

10. $8pq$: $p \equiv q \equiv 5$   (mod 8) (Corollary 9.8);

11. $8pq$: $p \equiv q + 4$   (mod 8), $\left(\frac{p}{q}\right) = -1$ (Corollary 9.8).

The status of the Pell equation remained fixed as above for more
than half a century, until Arnold Scholz extended the list of proved
cases by one, also explaining the earlier cases in his proof which used the
then-newly-developed results of class field theory [SCHO35]. Scholz's
addition covered the first open cases.

**Theorem 9.11.**

a) *Let* $\Delta = pq$ *be a discriminant such that* $p \equiv q \equiv +1$ (mod 4),
*and* $\left(\frac{p}{q}\right) = \left(\frac{q}{p}\right) = +1$. *If* $\left(\frac{p}{q}\right)_4 = -\left(\frac{q}{p}\right)_4$, *then (9.1) is not solvable
and* $4 \parallel h$.

b) *Let* $\Delta = pq$ *be a discriminant such that* $p \equiv q \equiv +1$ (mod 4),
*and* $\left(\frac{p}{q}\right) = \left(\frac{q}{p}\right) = +1$. *If* $\left(\frac{p}{q}\right)_4 = \left(\frac{q}{p}\right)_4 = +1$, *then (9.1) can be
either solvable or not solvable and* $8 \parallel h$.

c) *Let* $\Delta = 8p$ *be a discriminant such that* $p \equiv +1$ (mod 8). *If*
$p \equiv +1$ (mod 16) *and* $\left(\frac{2}{p}\right)_4 = -1$, *then (9.1) is not solvable and*
$4 \parallel h$.

*d) Let $\Delta = 8p$ be a discriminant such that $p \equiv +1 \pmod 8$. If $p \equiv +9 \pmod{16}$ and $\left(\frac{2}{p}\right)_4 = +1$, then (9.1) can be either solvable or not solvable and $8 \parallel h$.*

The even and odd cases of Theorem 9.11 can be united if we make the assumption that $\left(\frac{p}{8}\right)_4$ is $+1$ or $-1$ according as $p$ is congruent to 1 or 9 modulo 16. By this convention, Theorem 9.10 can be subsumed into Theorem 9.9.

## 9.2  Modern Results

As we have seen, the structure of the 2-Sylow subgroup is defined in part by the distribution of the ambiguous forms in the classes and genera, and this subsumes the question of solutions of the negative Pell equation (9.1). That is, (9.1) is solvable if and only if the reduced form $(-1, *, *)$ lies in the principal cycle. This is true if and only if *each* reduced ambiguous form $(a, ka, c)$ is equivalent to its "negative" $(-a, ka, -c)$, and this in turn is true if and only if every form $ax^2 - (k^2a - 4c)y^2$ represents $-4a$ and not some other divisor of the discriminant.

For a more careful examination of the issue of representation, we need further terminology. The forms $(a, ka, c)$ are the *ambiguous* forms. We also have used the word ambiguous for forms $(a, b, a)$, noting that such a form is equivalent "in one step" to a form $(b+2a, b+2a, a)$. Following Pall [PALL69] we shall call primitive forms $(a, 0, c)$ or $(a, a, c)$ *ancipital*. As will be seen shortly, ancipital and ambiguous forms correspond; we use the ancipital forms to simplify the basic algebra involved, especially for positive even discriminant, since the ancipital forms expose as the third coefficient the "rest" of the factoring $4ac$ of the discriminant. Ancipital forms occur in pairs, with $(a, 0, c)$ and $(c, 0, a)$ occurring together and corresponding to a factoring $4ac = -\Delta$ of the discriminant $\Delta$, and $(a, a, c)$ and $(4a-c, 4a-c, c)$ occurring together and corresponding to a factoring $a(4c - a) = -\Delta$. The two forms of each kind of pair are clearly equivalent. Further, for discriminants not a perfect square, exactly one of $|a|$ or $|c|$ is less than $\sqrt{|\Delta/4|}$ in the first case, and exactly one of $|a|$ or $|4c - a|$ is less than $\sqrt{|\Delta|}$ in the second. The set $S_1$ of first coefficients $a$ of the ancipital forms $(a, 0, c)$ for which we have $|a| < \sqrt{|\Delta/4|}$ is called the set of *discriminantal divisors of the first kind.* The set $S_2$ of first coefficients $a$ of the ancip- ital forms $(a, a, c)$ for which we have $\mid a \mid < \sqrt{\mid \Delta \mid}$ is called the set of *discriminantal divisors of the second kind.* The two sets taken together will be called simply discriminantal divisors. The following theorem is cited by Pall and implicit in Gauss. We present only an informal proof.

**Theorem 9.12.** *For positive discriminant, an ambiguous class con- tains exactly two ancipital forms of positive first coefficient and repre- sents exactly two discriminantal divisors. For negative discriminant not equal to −3 or −4, an ambiguous class contains exactly one ancipital form and represents exactly one discriminantal divisor.*

In the easier case of negative discriminant, an ancipital form $(a, 0, c)$ with $a$ positive and a discriminantal divisor is necessarily reduced, as we have $0 < a < c$. An ancipital form $(a, a, c)$ may not be reduced, as we may have $0 < a < \sqrt{\mid \Delta \mid}$ but not $0 < a < \sqrt{\mid \Delta/3 \mid}$, or equivalently, not have $a \leq c$. In this case, $(c, a - 2c, c)$ is reduced. As an example we may take $\Delta = -435$, with ancipital forms $(1, 1, 109)$, $(3, 3, 109)$, and $(5, 5, 109)$, which are reduced, and $(29, 29, 11)$, which is not, but is equivalent to the reduced form $(11, 7, 11)$.

In the case of positive discriminant, every ambiguous class contains two ambiguous forms $(a, ka, c)$. Each of these is equivalent to either $(a, 0, c')$ or $(a, a, c')$, according as $k$ is even or odd, and $a$ and $c'$ are of opposite sign. If the negative Pell equation is solvable, then each reduced cycle will contain pairs $(a, ka, c)$ and $(-a, ka, -c)$, with $\pm a$ the represented discriminantal divisors. If the negative Pell equation is not solvable, then each cycle will represent two distinct discriminantal divisors. It is this sort of representation which is used in Dirichlet's proofs earlier in this section.

It is clear that ancipital forms of the second kind $(a, a, c)$ with $a$ even can only occur for nonfundamental discriminants. It is easy to show, as Pall does, that forms $(a, 0, c)$ and $(a, a, c')$ can be equivalent only if we have $128 \mid \Delta$ and $32 \mid a$. Pall gives $(32, 0, -41)$ and $(32, 32, -33)$ of discriminant 5248 as an example of when this phenomenon occurs.

Another easy theorem is proved in [BROW71a], essentially by considering primitivity of forms.

**Theorem 9.13.** *Let* $\Delta = 2^t p_1^{a_1} \ldots p_r^{a_r}$ *be a positive nonsquare discriminant, in which the* $p_i$ *are distinct odd primes,* $n \geq 0$, *and* $a_i \geq 1$ *for each* $i$. *Let* $m$ *be a discriminantal divisor of* $\Delta$.

   *i) If* $p_i \mid m$, *then* $p_i^{a_i} \parallel m$.

   *ii) If* $m$ *is even, then* $2 \parallel m$ *if* $n = 2$ *or 3, and either* $4 \parallel m$ *or* $2^{n-2} \parallel m$ *if* $n > 3$.

   *iii) If* $n = 2$ *and* $m$ *is even, then* $\Delta \equiv 12 \pmod{16}$.

We now prove the theorem which, although unfortunately complicated to state, explicitly connects the representations of the discriminantal divisors by the principal forms of odd discriminants $\Delta$ and derived discriminants $4\Delta$. To simplify somewhat the terminology, we shall say that a form $(a, b, c)$ represents $m$ with $(\alpha, \gamma)$, or equivalently that $m$ is represented by $(a, b, c)$ with $(\alpha, \gamma)$, if there exist integers $\alpha$ and $\gamma$ such that $a\alpha^2 + b\alpha\gamma + c\gamma^2 = m$. We shall also say that the matrix $M$ takes a form $f = (a, b, c)$ to a form $f' = (a', b', c')$ if, under the transformation

$$\begin{pmatrix} x \\ y \end{pmatrix} = M \begin{pmatrix} x' \\ y' \end{pmatrix},$$

we have

$$ax^2 + bxy + cy^2 = a'x'^2 + b'x'y' + c'y'^2$$

We shall write this relationship as

$$M : (a,\ b,\ c) \to (a',\ b',\ c')$$

**Theorem 9.14.** *Let $\Delta$ be an odd positive discriminant. If $\Delta = mn$ is a factoring of $\Delta$ for which $(1,\ 1,\ (1-\Delta)/4) \sim (m,\ m,\ (m-n)/4)$, then the following are true.*

a) *There exists a primitive representation of $m$ by $(1,\ 1,\ (1-\Delta)/4)$ with $(\alpha, \gamma)$ for which $\gamma$ is even.*

b) *The forms $(1,\ 0,\ -\Delta)$ and $(m,\ 0,\ -n)$ of discriminant $4\Delta$ are equivalent. Specifically, if $(1,\ 1,\ (1-\Delta)/4)$ represents $m$ with $(\alpha, \gamma)$ with $\gamma$ even, and if we choose $\beta$ and $\delta$ so that $\begin{pmatrix} \alpha & \beta \\ \gamma & \delta \end{pmatrix}$ takes $(1,\ 1,\ (1-\Delta)/4)$ to $(m,\ m,\ (m-n)/4)$, then the matrix*

$$\begin{pmatrix} \alpha + \gamma/2 & -\alpha + 2\beta - \gamma/2 + \delta \\ \gamma/2 & -\gamma/2 + \delta \end{pmatrix}$$

*takes $(1,\ 0,\ -\Delta)$ to the form $(m,\ 0,\ -n)$, so that $(1,\ 0,\ -\Delta)$ primitively represents $m$ with $(\alpha + \gamma/2, \gamma/2)$.*

**Proof.** The assumption is that $(1,\ 1,\ (1-\Delta)/4)$ primitively represents $m$ and $-n$, both factors of the discriminant. We have stated and will prove the theorem without regard to the magnitude or the signs of $m$ and $n$. The tedious part of this proof consists of showing that in all cases we may find a representation for which $\gamma$ is even. We put that off until later and for the moment assume that $\gamma$ is even.

We observe that

$$\begin{pmatrix} \alpha & \beta \\ \gamma & \delta \end{pmatrix} \begin{pmatrix} 1 & 0 \\ 0 & 2 \end{pmatrix} = \begin{pmatrix} 1 & 0 \\ 0 & 2 \end{pmatrix} \begin{pmatrix} \alpha & 2\beta \\ \gamma/2 & \delta \end{pmatrix}$$

and that the last matrix is in the modular group if $\gamma$ is even. We consider the forms as $2 \times 2$ matrices, and form equivalence as matrix equivalence: If $f_1$ and $f_2$ are forms, and if $F_1$ and $F_2$ are the matrices which correspond to $f_1$ and $f_2$, then the matrix $M$ in the modular group takes $f_1$ to $f_2$ if and only if as matrices we have $F_2 = M^T F_1 M$.

Given $(\alpha, \gamma)$ with $\gamma$ even by which the principal form represents $m$, we may find integers $\beta$ and $\delta$ so that $\alpha\delta - \beta\gamma = 1$ and

$$\begin{pmatrix} m & m/2 \\ m/2 & (m-n)/4 \end{pmatrix} = \begin{pmatrix} \alpha & \gamma \\ \beta & \delta \end{pmatrix} \begin{pmatrix} 1 & 1/2 \\ 1/2 & (1-\Delta)/4 \end{pmatrix} \begin{pmatrix} \alpha & \beta \\ \gamma & \delta \end{pmatrix}$$

We now derive both sides with

$$\begin{pmatrix} 1 & 0 \\ 0 & 2 \end{pmatrix}$$

to obtain forms of discriminant $4\Delta$:

$$\begin{aligned}
\begin{pmatrix} m & m \\ m & m-n \end{pmatrix} &= \begin{pmatrix} 1 & 0 \\ 0 & 2 \end{pmatrix} \begin{pmatrix} m & m/2 \\ m/2 & (m-n/4) \end{pmatrix} \begin{pmatrix} 1 & 0 \\ 0 & 2 \end{pmatrix} \\
&= \begin{pmatrix} 1 & 0 \\ 0 & 2 \end{pmatrix} \begin{pmatrix} \alpha & \gamma \\ \beta & \delta \end{pmatrix} \begin{pmatrix} 1 & 1/2 \\ 1/2 & (1-\Delta)/4 \end{pmatrix} \begin{pmatrix} \alpha & \beta \\ \gamma & \delta \end{pmatrix} \begin{pmatrix} 1 & 0 \\ 0 & 2 \end{pmatrix} \\
&= \begin{pmatrix} \alpha & \gamma/2 \\ 2\beta & \delta \end{pmatrix} \begin{pmatrix} 1 & 0 \\ 0 & 2 \end{pmatrix} \begin{pmatrix} 1 & 1/2 \\ 1/2 & (1-\Delta)/4 \end{pmatrix} \begin{pmatrix} 1 & 0 \\ 0 & 2 \end{pmatrix} \begin{pmatrix} \alpha \\ \gamma/2 \end{pmatrix} \\
&= \begin{pmatrix} \alpha & \gamma/2 \\ 2\beta & \delta \end{pmatrix} \begin{pmatrix} 1 & 1 \\ 1 & 1-\Delta \end{pmatrix} \begin{pmatrix} \alpha & 2\beta \\ \gamma/2 & \delta \end{pmatrix}
\end{aligned}$$

With our assumption that $\gamma$ is even, this provides a proper equivalence of $(1, 2, 1-\Delta)$ and $(m, 2m, m-n)$. We now apply the inverse of the group generator $S$ to produce ancipital forms.

$$
\begin{aligned}
\begin{pmatrix} m & 0 \\ 0 & -n \end{pmatrix} &= \begin{pmatrix} 1 & 0 \\ -1 & 1 \end{pmatrix} \begin{pmatrix} m & m \\ m & m-n \end{pmatrix} \begin{pmatrix} 1 & -1 \\ 0 & 1 \end{pmatrix} \\
&= \begin{pmatrix} 1 & -1 \\ 0 & 1 \end{pmatrix} \begin{pmatrix} \alpha & \gamma/2 \\ 2\beta & \delta \end{pmatrix} \begin{pmatrix} 1 & 1 \\ 1 & 1-\Delta \end{pmatrix} \begin{pmatrix} \alpha & 2\beta \\ \gamma/2 & \delta \end{pmatrix} \begin{pmatrix} 1 & -1 \\ 0 & 1 \end{pmatrix} \\
&= \begin{pmatrix} \alpha & \gamma/2 \\ -\alpha+2\beta & -\gamma/2+\delta \end{pmatrix} \begin{pmatrix} 1 & 1 \\ 1 & 1-\Delta \end{pmatrix} \begin{pmatrix} \alpha & -\alpha+2\beta \\ \gamma/2 & -\gamma/2+\delta \end{pmatrix} \\
&= \begin{pmatrix} \alpha & \gamma/2 \\ -\alpha+2\beta & -\gamma/2+\delta \end{pmatrix} \begin{pmatrix} 1 & 0 \\ 1 & 1 \end{pmatrix} \begin{pmatrix} 1 & 0 \\ 0 & -\Delta \end{pmatrix} \\
&\quad \cdot \begin{pmatrix} 1 & 1 \\ 0 & 1 \end{pmatrix} \begin{pmatrix} \alpha & -\alpha+2\beta \\ \gamma/2 & -\gamma/2+\delta \end{pmatrix} \\
&= \begin{pmatrix} \alpha+\gamma/2 & \gamma/2 \\ -\alpha+2\beta-\gamma/2+\delta & -\gamma/2+\delta \end{pmatrix} \begin{pmatrix} 1 & 0 \\ 0 & -\Delta \end{pmatrix} \\
&\quad \cdot \begin{pmatrix} \alpha+\gamma/2 & -\alpha+2\beta-\gamma/2+\delta \\ \gamma/2 & -\gamma/2+\delta \end{pmatrix}
\end{aligned}
$$

Except for our assumption that $\gamma$ is even, the theorem is proved.

For brevity, we shall call the representations by $(t, u)$ (odd, odd), (even, odd), or (odd, even) according to the reduction of the pair modulo 2. (The (even, even) possibility cannot occur for primitive representations, so we need not consider it.) We observe that $(\alpha, \gamma)$ represents $m_1$ if and only if $(\alpha+\gamma, \gamma)$ does, and that if one of these is an (odd, odd) representation, the other is (even, odd). Thus we may always choose a representation which is either (even, odd) or (odd, even). We must show that the former are also unnecessary.

There are three main cases.

A. If $\Delta$ is congruent to 1 modulo 8, then the principal form represents odd numbers exactly when the representations are (odd, even).

B. If $\Delta$ is congruent to 5 modulo 8, then the fundamental solution of the Pell equation $T^2 - \Delta U^2 = 4$ could either have $T$ and $U$ both odd or $T = 2t$ and $U = 8u$ with $t = 8\tau \pm 1$ for an integer $\tau$.

1. If $T$ and $U$ are both odd, then we can convert an (even, odd) representation into an (odd, even) one.

2. If the fundamental solution (and hence all solutions) of the Pell equation have $T$ and $U$ both even, then we can explicitly construct (odd, even) representations of $m$.

In case B.1 we choose the signs of $T$ and $U$ so that we have $T + U \equiv 0$ (mod 4). Given an (even, odd) representation $(\alpha, \gamma)$, we have

$$\left( \frac{(2\alpha + \gamma) + \gamma \sqrt{\Delta}}{2} \right) \left( \frac{(2\alpha + \gamma) - \gamma \sqrt{\Delta}}{2} \right) = m,$$

with the coefficients of the rational and irrational parts being odd integers. Multiplying by

$$\left( \frac{T + U\sqrt{\Delta}}{2} \right) \left( \frac{T - U\sqrt{\Delta}}{2} \right) = 1$$

we have

$$\left( \frac{(2\alpha + \gamma)T + \gamma U\Delta + [\gamma T + (2\alpha + \gamma)U]\sqrt{\Delta}}{4} \right)$$
$$\cdot \left( \frac{(2\alpha + \gamma)T + \gamma U\Delta - [\gamma T + (2\alpha + \gamma)U]\sqrt{\Delta}}{4} \right) = m.$$

We have an (odd, even) representation if and only if the 4 in the denominator divides the coefficients of both the rational and irrational parts in each factor. Given our choice of signs for $T$ and $U$ and the fact that $\alpha$ is even, this must be true.

In case B.2, we have yet four subcases.

i) $\tau \equiv 1 \pmod 8$ and $\tau(4\tau + 1) = u^2 \Delta$.

ii) $\tau \equiv 7 \pmod 8$ and $\tau(4\tau - 1) = u^2 \Delta$.

iii) $\tau = 2^{2e} \tau_0$, $u = 2^e u_0$, $e \geq 1$, $u_0$ odd, $\tau_0 \equiv 5 \pmod 8$ and $\tau_0(2^{2e+2}\tau_0 + 1) = u_0^2 \Delta$.

iv) $\tau = 2^{2e} \tau_0$, $u = 2^e u_0$, $e \geq 1$, $u_0$ odd, $\tau_0 \equiv 3 \pmod 8$ and $\tau_0(2^{2e+2}\tau_0 - 1) = u_0^2 \Delta$.

In the first case, we let $\tau = u_1^2 d_1$ and $4\tau + 1 = u_2^2 d_2$ so that $\Delta = d_1 d_2$. The principal form of discriminant $\Delta$ represents $-d_1$ with the (odd, even) pair $(2\tau/u_1 - u_2, 2u_2)$ and represents $d_2$ with the (odd, even) pair $((4\tau + 1)/u_2 - 2u_1, 4u_1)$. By the congruence conditions, neither $-d_1$ nor $d_2$ is congruent to 1 modulo 8, and so neither can be 1. Thus, these are representations of nontrivial divisors of the discriminant.

The proofs for the other three cases are similar. In case ii, $(2\tau/u_1 - u_2, 2u_2)$ and $((4\tau - 1)/u_2 - 2u_1, 4u_1)$ represent $d_1$ and $-d_2$, respectively. In cases iii and iv, $(2^{e+1}\tau_0/u_1 - u_2, 2u_2)$ and $((2^{2e+2}\tau_0 \pm 1)/u_2 - 2u_1, 4u_1)$ represent $\mp d_1$ and $\pm d_2$, respectively, with the $\pm$ signs linked. The only minor problem occurs in case iii. If we let $\tau_0 = u_1^2 d_1$ and $2^{2e+2}\tau_0 + 1 = u_2^2 d_2$, then we could have $d_1 = \Delta$ and $d_2 = 1$. However, if this is true, then we have $(2u_2)^2 - (2^{e+2}u_1)^2 = 4$, which contradicts our choice of $(T, U)$ as the fundamental solution. Theorem 9.14 is proved.

**Corollary 9.15.** *Let $\Delta$ be an odd positive discriminant. The discriminantal divisors for discriminants $\Delta$ and $4\Delta$ are the same.*

At several points in this chapter we have skirted a general result which we will now formally discuss. For example, for the principal form of discriminant $\Delta$ to represent $-1$, it was seen to be necessary that all the odd prime divisors of $\Delta$ be congruent to 1 modulo 4, that is, that all prime divisors be representable by the (unique) class of discriminant $-4$. This is in fact generally true with some careful modifications. That is, representations of $m$ by the principal form of discriminant $\Delta = mn$ imply and are implied by facts regarding both the existence and the actual integers involved in the representation of $\Delta$ by forms of discriminant $2^k m$ for an appropriate $k$. Many of these results have been formally stated by Pall [PALL69] and by Brown [BROW71a], but they were more or less known long before.

We shall first cite a result of Cantor [CANT] which can be proved by direct computation.

**Theorem 9.16.** *Let $f_1 = (a_1,\ b_1,\ c_1)$ and $f_2 = (a_2,\ b_2,\ c_2)$ be forms of arbitrary discriminant. If, under*

$$\begin{pmatrix} X \\ Y \end{pmatrix} = \begin{pmatrix} \alpha & \beta \\ \gamma & \delta \end{pmatrix} \begin{pmatrix} x \\ y \end{pmatrix}$$

*we have that*

$$a_1 X^2 + b_1 XY + c_1 Y^2 = t_1 x^2 + u_1 xy + v_1 y^2$$
$$a_2 X^2 + b_2 XY + c_2 Y^2 = t_2 x^2 + u_2 xy + v_2 y^2$$

*then*

$$a_1 c_2 + a_2 c_1 - \frac{b_1 b_2}{2} = \left( t_1 v_2 + t_2 v_1 - \frac{u_1 u_2}{2} \right)(\alpha\delta - \beta\gamma).$$

We can make use of "Cantor's method" in proving the following theorem.

**Theorem 9.17.** *Let $\Delta$ be an odd discriminant. Then $(1,\ 1,\ \frac{1-\Delta}{4})$ represents $-1$ if and only if there exists a form $(-a,\ b,\ a)$ equivalent to $(1,\ 1,\ \frac{1-\Delta}{4})$. That is, the principal form of discriminant $\Delta$ represents $-1$ if and only if the principal form of discriminant $-16$, $(1,\ 0,\ 4)$, represents $\Delta$, $b^2 + 4a^2 = \Delta$, in such a way that both $a$ and $b$ are represented by the principal form of discriminant $\Delta$ and the related form $(-a,\ b,\ a)$ is in the principal class. This necessarily implies that $(\frac{a}{p}) = (\frac{b}{p}) = (\frac{-1}{p}) = +1$ for all primes $p$ dividing $\Delta$.*

**Proof.** We may by Theorem 9.14 choose the representation of $-1$ as $\alpha^2 + \alpha\gamma + \gamma^2(\frac{1-\Delta}{4}) = -1$ with $\gamma$ even. Then we have $(\alpha + \gamma/2)^2 - \Delta(\gamma/2)^2 = -1$ and there exists a primitive form $(\Delta\gamma/2,\ 2\alpha + \gamma,\ \gamma/2)$ of discriminant $-4$. Since $-4$ has class number 1, there exists a transformation matrix $M$ such that

$$M : (1,\ 0,\ 1) \to (\Delta\gamma/2,\ 2\alpha + \gamma,\ \gamma/2)$$

Consequently we have

$$M^T : (1,\ 0,\ -\Delta) \to (A,\ 2B,\ C)$$

$$M : (C,\ -2B,\ A) \to (-\Delta,\ 0,\ 1)$$

and by Cantor's theorem $A + C = 0$. We thus have $a$ and $b$ such that

$(1,\ 0,\ -\Delta) \sim (-b,\ -4a,\ b) \sim (-b,\ -4a - 2b,\ -4a)$. This last form is derived from $(-b,\ -2a - b,\ -a) \sim (-a,\ b,\ a)$, which is thus in the principal class of discriminant $\Delta$. This argument being reversible, the theorem is proved.

A second and direct proof, by a method which does not generalize, is this. Assuming that the principal form of discriminant $\Delta$ represents $-1$, we cycle forward from $(1,\ t,\ -u)$ and backward from $(u,\ t,\ -1)$:

$$(1,\ t,\ -u) \sim (-u,\ t',\ u') \sim (u',\ t'',\ -u'') \sim \cdots$$
$$\cdots \sim (u'',\ t'',\ -u') \sim (-u',\ t',\ u) \sim (u,\ t,\ -1)$$

Halfway between these two, or one-quarter of the way into the full cycle, we must come upon a form $(-a,\ b,\ a)$. For the converse, we note that if

$$M : (1,\ t,\ -u) \rightarrow (-a,\ b,\ a)$$

then we have

$$M^T : (-a,\ b,\ a) \rightarrow (u,\ t,\ -1)$$

and

$$M \cdot M^T : (1,\ t,\ u) \rightarrow (u,\ t,\ -1).$$

The general results of this type are complicated but not difficult. Rather than deal with the minutiae of the many cases, we shall prove only the most elegant case.

**Theorem 9.18.** *Let $4\Delta = 4mn \equiv 4 \pmod 8$ be a fundamental discriminant, and assume that $2m$ and $-2n$ are the discriminantal divisors represented by the principal form of discriminant $4\Delta$. Then the following are true:*

a) *Modulo each prime $p$ dividing $m$, we have $\left(\frac{-n}{p}\right) = \left(\frac{2}{p}\right)$.*

b) *There exists a form in the genus of $(2,\ 0,\ -m)$ of discriminant $8m$ which represents $-n$.*

c) *There exists a form in the principal genus of discriminant $8m$ which represents $-2n$.*

d) *The form in part b is $(2, 0, -m)$, that is, we have a representation $2a^2 - mb^2 = -n$, if and only if the form in part c is the principal form, representing $-2n$ as $(2a)^2 - 2mb^2 = -2n$, which is true if and only if there exist ambiguous forms $(2ma, 2mb, a)$ and $(ma, 2mb, 2a)$ of discriminant $4\Delta$, with a and b odd.*

e) *From the assumed representation $m^2 r^2 - \Delta s^2 = 2m$ of $2m$ we infer the existence of a primitive form $(\Delta s, 2mr, s)$ of discriminant $8m$. This form is equivalent to $(1, 0, -2m)$ if and only if $(1, 0, -\Delta)$ is equivalent to a form $(2ma, 2mb, a)$. The form $(\Delta s, 2mr, s)$ is equivalent to $(2, 0, -m)$ if and only if $(1, 0, -\Delta)$ is equivalent to a form $(ma, 2mb, 2a)$.*

f) *Parts a through e remain true if m and $-n$ are interchanged and the form $(\Delta s, 2mr, s)$ is replaced by the form $(\Delta r, 2ms, r)$.*

**Proof.** We begin with the remark that fundamental discriminants not divisible by 8 can represent even discriminantal divisors only in the case of discriminants $4\Delta = 4mn$, with $2m$ and $-2n$ being represented, and the representations are $(mr)^2 - \Delta s^2 = 2m$ and $(ns)^2 - \Delta r^2 = -2n$. In this, $r$ and $s$ are both odd, so we have $mr^2 - ns^2 = 2$. Part a is thus clearly true. The discriminant $8m$ has the additional character $\psi$ or $\chi\psi$ according as $m$ is 1 or 3 modulo 4. If $m$ is 1 modulo 8, then $-n$ is also 1, and $\psi(-m) = \psi(-n)$, so that $-n$ is represented by some form in the genus of $(2, 0, -m)$. The other cases are similar, and parts b and c are proved. Part d is self-evident except for the fact that $(2ma, 2mb, a)$ and $(ma, 2mb, 2a)$ are actually ambiguous. But, for example, the square of $(2ma,\ 2mb,\ a)$ is equivalent to $(2m,\ 2mb,\ a^2)$, which is principal. For part e, we apply Cantor's method to see that, with $i$ equal either to 0 or to 1, we have

$$M : (2^i,\ 0,\ -2^{1-i}m) \to (\Delta s,\ 2mr,\ s)$$

if and only if

$$M^T : (1,\ 0,\ -\Delta) \to M : (2^{1-i},\ 2mb,\ 2^i a).$$

## 9.3   Reciprocity Laws

From the preceding, especially the theorems of Dirichlet and Scholz, it is seen that solutions of the negative Pell equation are connected to power residue conditions among the primes dividing the discriminant. The case of discriminants with two or three prime divisors has been most thoroughly studied. Klaus Burde [BURD69] opened a fruitful line of inquiry with the following theorem, which we shall not prove.

**Theorem 9.19.** *Let $p$ and $q$ be primes, both congruent to 1 modulo 4, for which $\left(\frac{p}{q}\right) = +1$. If we write $p = a^2 + b^2$ and $q = c^2 + d^2$ with $a$ and $c$ odd and $b$ and $d$ even and $ab$ and $cd$ both positive, then*

$$\left(\frac{p}{q}\right)_4 \left(\frac{q}{p}\right)_4 = (-1)^{\frac{p-1}{4}} \left(\frac{ad - bc}{p}\right)$$

Burde's "rational reciprocity law" was used by Brown [BROW72a] to prove Scholz's theorem (our Theorem 9.11) by essentially elementary and constructive means.

Thus far we have considered only residue conditions on rational integers, but this is only a small part of the story. If primes $p$ and $q$ are both congruent to 1 modulo 4 and are mutual quadratic residues, then we may regard $\sqrt{q}$ and $\sqrt{p}$ as integers modulo $p$ and $q$, respectively. Thus the usual Legendre symbol $\left(\frac{\sqrt{q}}{p}\right)$ is well defined in terms of the rational integers. Part of Scholz's original result is the following theorem.

**Theorem 9.20.** *Let $p$ and $q$ be primes, both congruent to 1 modulo 4, for which $\left(\frac{p}{q}\right) = +1$. If $\varepsilon_p$ and $\varepsilon_q$ are the fundamental units of the* *quadratic fields of discriminants p and q, respectively, then*

$$\left(\frac{p}{q}\right)_4 \left(\frac{q}{p}\right)_4 = \left(\frac{\varepsilon_q}{p}\right) = \left(\frac{\varepsilon_p}{q}\right)$$

This has led to substantial interest in the quadratic and later quartic character of quadratic units.

# 9.4   Special References for Chapter 9

Pierre Barrucand and Harvey Cohn, "Note on primes of type $x^2 + 32y^2$, class number, and residuacity," *Crelle*, v. 238, 1969, 67-70.

Helmut Bauer, "Zur Berechnung der 2-Klassenzahl der quadratischen Zahlkörper mit genau zwei verschiedenen Diskriminantenprimteilern," *Crelle*, v. 248, 1971, 43-46.

Helmut Bauer, "Die 2-Klassenzahlen spezieller quadratischer Zahlkörper," *Crelle*, v. 252, 1972, 79-81.

Jacob A. Brandler, "Residuacity properties of real quadratic units," *Journal of Number Theory*, v. 5, 1973, 271-286.

Ezra Brown, "Representation of discriminantal divisors by binary quadratic forms," *Journal of Number Theory*, v. 3, 1971, 213-225.

Ezra Brown, "A theorem on biquadratic reciprocity," *Proceedings of the American Mathematical Society*, v. 30, 1971, 220-222.

Ezra Brown, "Binary quadratic forms of determinant $-pq$," *Journal of Number Theory*, v. 4, 1972, 408-410.

Ezra Brown, "Quadratic forms and biquadratic reciprocity," *Crelle*, v. 253, 1972, 214-220.

Ezra Brown, "Biquadratic reciprocity laws," *Proceedings of the American Mathematical Society*, v. 37, 1973, 374-376.

Ezra Brown, "Class numbers of complex quadratic fields," *Journal of Number Theory*, v. 6, 1974, 185-191.

Ezra Brown, "The class number and fundamental unit of $\mathbf{Q}(\sqrt{2p})$, for $p \equiv 1 \pmod{16}$ a prime," *Journal of Number Theory*, v. 16, 1983, 95-99.

Ezra Brown and Charles J. Parry, "Class numbers of imaginary

quadratic fields having exactly three discriminantal divisors,"
*Crelle*, v. 260, 1973, 31-34.

J. Bucher, "Neues über die Pell'sche Gleichung," *Naturforschende Gesellschaft Mitteilungen Luzern*, v. 14, 1943, 1-18.

Klaus Burde, "Ein rationales biquadratisches Reziprozitätsgesetz," *Crelle*, v. 235, 1969, 175-184.

Duncan A. Buell and Kenneth S. Williams, "An octic reciprocity law of Scholz type," *Proceedings of the American Mathematical Society*, v. 77, 1979, 315-318.

Duncan A. Buell, Philip A. Leonard, and Kenneth S. Williams, "Note on the quadratic character of a quadratic unit," *Pacific Journal of Mathematics*, v. 92, 1981, 35-38.

P. Epstein, "Zur Auflösbarkeit der Gleichung $x^2 - Dy^2 = -1$," *Crelle*, v. 171, 1934, 243-252.

Franz Halter-Koch, "An Artin character and representations of primes by binary quadratic forms III," *manuscripta mathematica*, v. 51, 1985, 163-169.

Franz Halter-Koch, Pierre Kaplan, and Kenneth S. Williams, "An Artin character and representations of primes by binary quadratic forms II," *manuscripta mathematica*, v. 37, 1982, 357-381.

Helmut Hasse, "Über die Teilbarkeit durch $2^3$ der Klassenzahl der quadratichen Zahlkörper mit genau zwei verschiedenen Diskriminantenprimteilern," *Mathematische Nachrichten*, v. 46, 1970, 61-70.

Helmut Hasse, "Über die Klassenzahl des Körpers $P(\sqrt{-2p})$ mit einer Primzahl $p \neq 2$," *Journal of Number Theory*, v. 1, 1969, 231-234.

Pierre Kaplan, "Divisibilité par 8 du nombre des classes des corps quadratiques dont le 2-groupe des classes est cyclique, et réciprocité biquadratique," *Journal of the Mathematical Society of Japan*, v. 25, 1973, 596-608.

Pierre Kaplan, "Sur le 2-groupe des classes d'ideaux des corps quadratiques," *Crelle*, v. 283/284, 1976, 313-363.

Pierre Kaplan and Kenneth S. Williams, "An Artin character and representations of primes by binary quadratic forms," *manuscripta mathematica*, v. 33, 1981, 339-356.

Emma Lehmer, "Criteria for cubic and quartic residuacity," *Mathematika*, v. 5, 1958, 20-29.

Emma Lehmer, "On the quadratic character of some quadratic surds," *Crelle*, v. 220, 1971, 42-48.

Emma Lehmer, "On some special quartic reciprocity laws," *Acta Arithmetica*, v. 21, 1972, 367-377.

Emma Lehmer, "On the quartic character of quadratic units," *Crelle*, v. 268/269, 1974, 294-301.

Emma Lehmer, "Rational reciprocity laws," *American Mathematical Monthly*, v. 85, 1978, 467-472.

Philip A. Leonard and Kenneth S. Williams, "The quartic characters of certain quadratic units," *Journal of Number Theory*, v. 12, 1980, 106-109.

Philip A. Leonard and Kenneth S. Williams, "A representation problem involving binary quadratic forms," *Archiv der Mathematik*, v. 36, 1981, 53-56.

Philip A. Leonard and Kenneth S. Williams, "The quadratic and quartic characters of certain quadratic units I," *Pacific Journal of Mathematics*, v. 71, 1977, 101-106.

Philip A. Leonard and Kenneth S. Williams, "The quadratic and quartic characters of certain quadratic units II," *Rocky Mountain Journal of Mathematics*, v. 9, 1979, 683-692.

Patrick Morton, "On Rédei's theory of the Pell equation," *Crelle*, v. 307/308, 1979, 373-398.

L. Rédei, "Arithmetischer Beweis des Satzes über die Anzahl der durch 4 teilbaren Invarianten der absoluten Klassengruppe im quadratischen Zahlkörper," *Crelle*, v. 171, 1934, 55-60.

L. Rédei, "Eine obere Schranke der Anzahl der durch vier teilbaren Invarianten der absoluten Klassengruppe im quadratischen Zahlkörper," *Crelle*, v. 171, 1934, 61-64.

L. Rédei, "Über die Grundeinheit und die durch 8 teilbaren Invarianten der absoluten Klassengruppe im quadratischen Zahlkörper," *Crelle*, v. 171, 1934, 131-148.

L. Rédei and H. Reichardt, "Die Anzahl der durch 4 teilbaren Invarianten der Klassengruppe eines beliebigen quadratischen Zahlkörpers," *Crelle*, v. 170, 1933, 69-74.

H. Reichardt, "Über die 2-Klassengruppe gewisser quadratischer Zahlkörper," *Mathematische Nachrichten*, v. 46, 1970, 71-80.

H. C. Williams, "The quadratic character of a certain quadratic surd," *Utilitas Mathematica*, v. 5, 1974, 49-55.

Kenneth S. Williams, "A rational octic reciprocity law," *Pacific Journal of Mathematics*, v. 63, 1976, 563-570.

Kenneth S. Williams, "Note on Burde's rational biquadratic reciprocity law," *Canadian Mathematics Bulletin*, v. 20, 1977, 145-146.

Kenneth S. Williams, "Congruences modulo 8 for the class numbers of $\mathbf{Q}(\sqrt{\pm p})$, $p \equiv 3 \pmod 4$ a prime," *Journal of Number Theory*, v. 15, 1982, 182-198.

Kenneth S. Williams, "On Scholz's reciprocity law," *Proceedings of the American Mathematical Society*, v. 46, 1977, 45-46.

Kenneth S. Williams and James D. Currie, "Class numbers and biquadratic reciprocity," *Canadian Journal of Mathematics*, v. 34, 1982, 969-988.

# Chapter 10

# Factoring with Binary Quadratic Forms

## 10.1  Classical Methods

In the opening paragraphs of Article 329 of the *Disquisitiones*, Gauss, a master at calculating, writes [pp. 396-397],

> The problem of distinguishing prime numbers from composite numbers and of resolving the latter into their prime factors is known to be one of the most important and useful in arithmetic ... the dignity of the science itself seems to require that every possible means be explored for the solution of a problem so elegant and so celebrated.

One really needs no more suggestion than this to pursue this beguiling topic of factoring an integer into the product of its component primes. The earliest techniques for factoring an integer $N$ involve binary quadratic forms. According to Dickson's *History*, [DICK], Fermat (about1643) wrote that "An odd number not a square can be expressed as the difference of two squares in as many ways as it is the product of tation of $N$ as $x^2 - y^2$:

$$N = ab = \left(\frac{a+b}{2}\right)^2 - \left(\frac{a-b}{2}\right)^2.$$

Fermat used this technique to factor $100895598169 = 112303 \cdot 898423$.

The converse side of Fermat's statement is that $N$ is a prime if and only if $x = \frac{N+1}{2}$ and $y = \frac{N-1}{2}$ give the only representation of $N$ as $x^2 - y^2$.

On the one hand, finding such representations would seem to be an infinite task, since the equation is hyperbolic. However, recognizing that the two factors must be integers makes a search for representations finite, since $\mid x \mid \geq \mid y \mid +1$ leads to the necessary condition that $\mid y \mid \leq \frac{N-1}{2}$. Although it is finite, this bound is linear in $N$ and is thus not effective for large values of $N$.

Mersenne and then Euler were able to use sums of squares to advantage in factoring, as follows. For $N$ which are of the form $4k + 1$, $N$ can be written as a sum of squares $N = a^2 + b^2$ in only one way if $N$ is prime. Conversely, if $N$ is representable in two different ways as the sum of squares,

$$N = a^2 + b^2 = c^2 + d^2$$

then it can be seen that $N$ is composite, and a factoring arises from the algebraic identity

$$N = \frac{[(a-c)^2 + (b-d)^2][(a+c)^2 + (b-d)^2]}{4(b-d)^2} \qquad (10.1)$$

It is not difficult to recognize the rule of composition of forms in (10.1).

It was Euler who noted that the idoneal numbers were suitable for primality testing. Recall that an integer $N$ is representable by a form if and only if the congruence $b^2 = \Delta \pmod{4N}$ is solvable. If $N$ is prime, this congruence has exactly two solutions if it has any, and the solutions produce opposite forms. If $N$ has $k$ prime factors and the congruence is solvable, there are $2^k$ solutions. These may not all give rise to forms in the same class for an arbitrary discriminant $\Delta$, but in the idoneal discriminants, with one class per genus, if an integer is representable by some class of that discriminant, then all representations by forms of that discriminant are by forms in that unique class. This generalizes the remarks above about primality. An integer $N$ which is representable in only one way by a form of an idoneal discriminant must be prime.

Gauss, in the *Disquisitiones*, gave a factoring method based on representations of $N$ by the principal form of an idoneal discriminant. If we can represent $N$ in two ways by a form of discriminant $-4D$,

$$N = a^2 + Db^2 = c^2 + Dd^2$$

then, modulo $N$, we have

$$
\begin{aligned}
a^2 + Db^2 &\equiv c^2 + Dd^2 \equiv 0 \\
a^2d^2 + Db^2d^2 &\equiv b^2c^2 + Db^2d^2 \\
a^2d^2 - b^2c^2 &\equiv 0 \\
(ad + bc)(ad - bc) &\equiv 0
\end{aligned}
$$

Although we no longer have an identity that factors $N$, as with sums or differences of squares, we can hope to factor $N$ by taking $\gcd(ad + bc, N)$ and $\gcd(ad - bc, N)$.

These congruences are clearly not limited just to the idoneal discriminants; any discriminant for which one can obtain two representations by the same form will suffice. The utility of the idoneal discriminants is that one is guaranteed in this case that all representations are by the same form, so that the congruences are applicable.

An immediate advantage of the sums of squares techniques over the difference of squares technique is that the search is now bounded by a function of $\sqrt{N}$ and not $N$. That is, we have

$$|y| \le \sqrt{N/D}$$

Indeed, if there were infinitely many idoneal discriminants, one could hope to factor most, or even all, integers, with very little effort by choosing $D$ large enough to make the search limits on $y$ very small.

One way in which the search for representations for all three methods can be made simpler is by what is now known as Gaussian exclusion. The equations

$$x^2 - y^2 = N$$
$$x^2 + y^2 = N$$
$$x^2 + Dy^2 = N$$

can be considered as congruences modulo small primes and thus used to provide quadratic residue conditions on $x$ and $y$.

To illustrate all three methods, we will factor 22873. We first attempt to solve

$$x^2 - y^2 = 22873,$$

and will always assume that $x$ and $y$ are positive to eliminate needless notation. This equation, as a congruence modulo 8, is

$$x^2 \equiv y^2 + 1,$$

from which we infer that $x$ is odd and that $y$ is divisible by 4. Modulo 3, we also have

$$x^2 \equiv y^2 + 1,$$

from which we infer that $y$ is divisible by 3 and $x$ is not. Modulo 5, we find that $y$ must be congruent to 1 or to 4, and modulo 7, to 0, 2, or 5. The original 11436 possible values for $y$ are thus reduced to 163, and the solution

$$173^2 - 84^2 = 22873$$

is obtained after some effort.

In a similar vein we attempt to solve

$$x^2 + y^2 = 22873 \tag{10.2}$$

but now we need two different solutions. We note the advantage of a $\sqrt{N}$ approach immediately; we *start* with only 151 possible values for $y$. Working modulo 8 and modulo 3, we see that one of $x$ and $y$ is odd and the other divisible by 4, and that one is and the other is not divisible by 3. We may thus choose $x$ to be either of the form $12k$ or $6k + 3$ for suitable $k$ such that $x \leq 151$. As a congruence modulo 5, we find that (10.2) forces both $x$ and $y$ to be independently either 2 or 3. Solving the resulting congruences, the only possible values for $x$ are seen to be 12, 48, 72, 108, 132, 3, 33, 63, 93, 123, 30, 60, 90, 120, and 150. By trial and error we find that

$$22873 = 72^2 + 133^2 = 123^2 + 88^2.$$

If we examine the appropriate generic characters for idoneal discriminants, we find that 22873 is representable by $x^2 + 1848y^2$, the principal form of discriminant $-7392$, the idoneal discriminant largest in magnitude. With this large a discriminant, the only possible values of $y$ are 1, 2 or 3. It is then essentially trivial to determine that

$$22873 = 145^2 + 1848 \cdot 1^2 = 79^2 + 1848 \cdot 3^2.$$

With any of these methods, we can factor $22873 = 89 \cdot 257$.

We note in passing that there are factoring methods other than those we shall discuss here [BRIL83, BUEL87b, POME83, LENS, WAGS].

# 10.2 SQUFOF

In 1975 Daniel Shanks devised a clever and simple method for factoring, which he referred to as SQUFOF, for Square Factorization of Forms (or possibly Square FOrm Factorization). This is the first of three heuristic techniques we will discuss. Each of these techniques uses a different method for accomplishing the same goal–with $N$, the number be factored, as the discriminant (if $N$ cannot be a discriminant, we use a suitable multiple of $N$), we try to find the ambiguous forms of that discriminant $\Delta$. These lead to factors of $N$.

The ambiguous classes being the classes of order 2 in the group, one way to find factors of the discriminant is to find square roots of the principal class. We observe that if we have a form $(a^2, b, c)$, then it is the square of the form $(a, b, ac)$ up to common factors of $a$ and $b$. If $a$ and $b$ are both even there is the omnipresent annoyance of considering 2 as a special case, but any odd common factor must be exactly one of the nontrivial factors of the discriminant which we seek.

For negative discriminants, a reduced form $(a^2, b, c)$ cannot be principal, so although explicit square roots are possible the square roots are not ambiguous. For positive discriminants, however, we may expect to be able to find such forms in the principal cycle and take square roots, hoping thus to land in an ambiguous cycle and produce in the cycle the ambiguous form.

We illustrate SQUFOF with an example, trying to factor $N = 3292097$. Since $3292097 \equiv 1 \pmod 4$, we use $N$ as the discriminant. There are standard techniques for finding "small" factors of numbers, so it is normal for a "large" number as yet not factored but known to be composite to have two or three prime factors. Further, it is normal for positive discriminants for class numbers to be small, in general to have one class per genus. With these assumptions, using the Dirichlet formula, we estimate that the principal cycle has approximately $\sqrt{N}$ or about 1813 forms. In fact, the number is 1018. There is, of course, a second ambiguous form in the principal cycle and a naive way to factor $N$ would be to progress through the entire cycle to find that other form. This would take $\sqrt{N}$ steps, however, which is much too large. In this instance, the principal form is $(1, 1813, -1282)$ and the other ambiguous form is $(-1, 1813, 1282)$, so the effort would be futile after all.

Beginning with the principal form, then, we cycle forward through adjacent forms of the principal cycle. The 40th form we come to is $(676, 1161, -719)$. We take its square root to get $(26, 1161, -18694) \sim (26, 1785, -1018)$. We now cycle either *forward* from $(26, 1785, -1018)^{-1}$ to $(26, 1803, -397)$ and on through to $(-397, 1373, 886) \dots$ or *backward* from $(26, 1785, -1018)$ through $(-397, 1803, 26) \dots$. Either way, the 22nd form we come to is the ambiguous form $(-1097, 1097, 476)$. We factor $3292097 = 1097 \cdot 3001$.

The work of Shanks and of Williams [WILL85] on the infrastructure of real quadratic fields has shown that rather precise estimates can be given of the positions of forms in cycles relative to the ambiguous forms. Thus, if the form $(a^2, b, c)$ occurs $K$ forms into the principal cycle from the principal form, then its square root $(a, b, ac)$ when properly reduced occurs approximately $K/2$ forms *in the same direction* from an ambiguous form. The directional nature of this rule is what forces us to cycle backward from the square root or forward from its inverse. In our example, the cycle containing $(26, 1785, -1018)$ has 1070 forms; by choosing the direction properly we arrive at the ambiguous form after only 21, and not 514, steps.

As with many computational heuristics, there are small variations in this method which either reduce the computational overhead or increase the probability of success (or decrease the probability of failure). First, cycling from $(a, b, c)$ to the adjacent form $(c, b', c')$ requires computing $\delta = \left\lceil \frac{\sqrt{\Delta} + b}{2|c|} \right\rceil$. We can eliminate the 2 in the denominator, and thus eliminate one computation (the multiplication by 2, or addition of $c$ to itself, or a one-bit left shift of $c$) in *every* step by always using discriminant $4N$ and not $N$, algebraically removing the 2's throughout the algorithm and the computer programs, and taking care to avoid producing the useless ambiguous form $(2, *, *)$. If $N \equiv 3 \pmod 4$, of course, this is necessary to get a discriminant of forms.

Second, the success of SQUFOF in factoring depends on the existence of squares as lead coefficients in the cycle. If the cycle for discriminant $4N$ seems to be devoid of these, then one might also try discriminants $4kN$ for small $k$, keeping in mind that the ambiguous forms so produced might now only factor $k$ and not $N$.

A third issue is the avoidance of failure by avoiding square roots that produce the principal cycle and not a nonprincipal ambiguous cycle. To do this we look again at the infrastructure of the cycle. The lead coefficients of reduced forms are necessarily smaller in magnitude that $\sqrt{\Delta}$. If such a coefficient is in fact a perfect square $k^2$, then we have that $| k | \leq \Delta^{1/4}$. Further, for even discriminants, if the form $(k, *, *)$ or $(2k, *, *)$ appears $K$ steps into the principal cycle, then $(k^2, *, *)$ will appear about $2K$ steps into the same cycle. It is taking square roots of such forms that we wish to avoid, so we shall, in progressing through the principal cycle, store a "stop list" of lead coefficients that are smaller in magnitude than $2\Delta^{1/4}$.

For example, the 44th form in the cycle of (1, 1813, −1282) is (256, 1505, −1003), and the 22d form is (−16, 1793, 1207). Had we not been storing a stop list, and encountered the square 256 before the square 676, we might have been tempted to cycle backward from (−16, 1793, 1207), only to find that we had never left the principal cycle.

We point out a curiosity in the computational framework of SQUFOF. Riesel [RIES85, pp. 191-199] describes SQUFOF solely in terms of continued fractions. Thus, he points out that we can only take square roots, looking for a perfect square, every second step–the odd steps correspond to negative lead coefficients of forms. Also, he apparently does not notice, or at least makes no mention, in his example that he *changes discriminant* upon taking the square root. In his example he expands the continued fraction of $\sqrt{1000009}$, equivalent to cycling through forms of discriminant 4000036. When he comes to the point corresponding to the form (16, 1974, −1615), he takes what corresponds to the explicit square root (4, 1974, −6460). This form, however, is *imprimitive*. In the continued fraction calculation the factors of 2 are cancelled, and one can proceed unaware of the fact that the correspondence of forms has indeed changed.

It remains to analyze the running time of SQUFOF. As with all three class-group factoring methods, precise theorems are unavailable, since much depends on unproved assumptions. However, since the assumptions are probably either true or true to a first order approximation, we can make a reasonable judgement as to the probable running time.

We first estimate the length of the principal cycle to be $O(\sqrt{\Delta})$, with the constant term generally quite small. Certainly the length of the cycle is roughly equal to the logarithm of the regulator $\ln \varepsilon$ in the Dirichlet formula for the class number. The $L$-functions, assuming the Riemann hypotheses, are bounded by something like $\log \log \Delta$. And if we assume that class numbers are on the average quite small, for example that one-class-per-genus groups predominate, then we see that indeed, to a first-order approximation, the principal cycle is of length $O(\sqrt{\Delta})$.

Next, we observe that there are $\Delta^{1/4}$ integers $k$ such that reduced forms $(k^2, *, *)$ exist. All these forms must lie in the principal genus; we assume their distribution is otherwise random. Since the number of classes per genus is assumed to be of much smaller order than $\sqrt{\Delta}$, we may assume then that these forms are distributed at intervals of roughly $\Delta^{1/4}$ throughout the principal cycle. Ignoring all lower-order terms, we thus expect to cycle forward $O(\Delta^{1/4})$ steps to find a form $(k^2, *, *)$, and then backward half then many steps to find a factor of $\Delta$. Although significant in practice, the effect of the stop list is decidedly of lower order and can thus be ignored. We would appear, then, based on some unproved assumptions which are nonetheless likely to be true, to have a factoring method whose order is the fourth root of $N$.

A final comment regarding SQUFOF concerns implementation. All the coefficients are of size $\sqrt{N}$, and thus the arithmetic necessary for computer implementation need only be for operands of half as many bits as the number to be factored. This has a significant effect on computational overhead, although it does not appear in order-of-magnitude estimates. Second, and most elegant, a program for SQUFOF is very small and simple; the algorithm has even been programmed very successfully on pocket calculators.

# 10.3  CLASNO

The second class-group-related factoring method is also due to Shanks
and is called CLASNO [SHAN69]. In a paper which describes many of
the computations in class groups which are now standard, Shanks gave
the first example of a factoring method which was essentially provably
of order $N^{1/4}$.

In CLASNO, we work with negative discriminants $\Delta = -N$ or
$-4N$, obtaining ambiguous forms by first estimating and then exactly
determining the class number $h$, and then finding an ambiguous form.
The first step is to estimate the class number using the Dirichlet formula

$$h(\Delta) = \frac{\sqrt{-\Delta}}{\pi} \prod_p \left( \frac{p}{p - \left(\frac{\Delta}{p}\right)} \right)$$

where the infinite product runs over all primes $p$. With a reasonably
streamlined program for computing quadratic residue symbols, and
some care with regard to numerical accuracy, one can fairly readily
get a good estimate for $h$ by cutting off the product at some upper
bound. Shanks says in his paper that a bound of primes less than
132000 for discriminants of 20 to 30 decimal digits produces an estimate which is generally good to within 1 part in 1000. The precise rate
of convergence of the product is linked to unproved hypotheses about
the distribution of quadratic residues and nonresidues modulo the discriminant, thus linked to the size of the class number itself, and to the
Riemann Hypotheses for the $L$-functions.

Having obtained the estimate for the class number $h$ by analytic
means, we now compute the class number exactly by using the group
structure. We will generally know something about the class number.
For example, if the discriminant is known to be composite, the class

number must of course be even. If the discriminant is four times a composite number, the class number must be divisible by 4. We assume, therefore, that we know $b$ such that $h \equiv 0 \pmod{b}$. Importantly, as the computations progress, we may discover further congruence conditions on $h$ that limit the possible values even further. We denote by $H$ the integer nearest the analytic estimate which satisfies the congruence conditions.

In the process of estimating $h$, we computed $\left(\frac{\Delta}{p}\right)$ for all small primes $p$. This information can be used to provide an essentially unlimited supply of forms $f$ of prime lead coefficient. We choose forms $f$ and compute $f^H$ until we find a form for which $f^H$ is not the identity. Of course, if we do exponentiate a form and by chance get the group identity, we obtain further congruence conditions on the order of the group.

If the estimate $H$ for $h$ is good, then we have $f^H \sim f^{H-h} \sim f^{bt}$ for some small integer t. We compute, for $k = 1, \ldots, s$ all reduced forms $f^{bk}$ and store the leading and center coefficients. (We shall shortly describe how to determine an appropriate value of $s$.) Because of the group structure, we now actually have $f^{bk}$ for $k = 0, \pm 1, \ldots, \pm s$. We search for a match between $f^H$ and our stored list of forms. If $H$ is within a distance $bs$ of $h$, a match will be found, and we will have determined the exact class number.

If we fail at this point to find a match, then $H$ differs from $h$ by at least $b(s + 1)$. We compute $f^{H+2bs}$ and $f^{H-2bs}$, and search once again for a match between one of these forms and a form in our stored list. As described by Shanks, we lay out an interval of "baby steps" of size $bs$ about the identity and shift $f^H$ by "giant steps" of size $bs$ until we get a match and determine the exact class number.

We now can factor $h = 2^i h'$, with $h'$ odd, and find some form $f^{h'}$ not the identity. This form must be of order some power of 2, so successive squaring will produce an ambiguous form.

The question arises about the appropriate size of $s$. In doing the giant steps and baby steps we perform $s$ compositions to get the forms $f^{bk}$ and some number, $2r$, of compositions to get the forms $f^{H \pm 2jbs}$ before a match is found. To minimize the number $s+2r$ of compositions, we should first choose $2r$ equal to $s$.

CLASNO succeeds when one of the $2rbs + bs = bs^2 + bs$ forms $f^l$ for $l = H - b(2rs + s), H - b(2rs + s - 1), \ldots, H + b(2rs + s)$ is the identity of the group. Thus, the estimate $H$ must be within $bs^2 + bs$ of $h$ in absolute value. If we know from experience that the estimate is good within 1 part in $K$, we should choose $s$ roughly equal to $\sqrt{\frac{h}{bK}}$ to minimize the total number of compositions.

To estimate the running time of this algorithm requires first determining what operations to count. Shanks considers only the number of compositions in the class group. If we make the major, but acceptable, assumption that $h = O(\sqrt{\Delta})$ for negative discriminants, then the number of compositions is about $\log \sqrt{\Delta}$ (for computing $f^H$) plus $\log \sqrt{\Delta}$ (for squaring $f^{h'}$ to get an ambiguous forms) plus $2r + s$ or about $\Delta^{1/4}$ for the search for a match, giving in sum the $O(\Delta^{1/4})$.

The other computations should, however, at least be mentioned. Computing a Jacobi symbol is essentially as complex as computing a greatest common divisor, and thus is theoretically easier only by a constant factor than is composition. Shanks uses an absolute bound of about $10^5$ to get 0.1% accuracy in estimating class numbers for discriminants of size about $10^{20}$. If in fact it is necessary to use a bound of $\Delta^t$ to get estimates within a constant percentage, then the estimate for $h$ is nontrivial. Similarly, the search for a match, if done after sorting the list of forms, is of order $\log_2 \Delta^{1/4}$, but the cost of the sort is $\Delta^{1/4} \log \Delta^{1/4}$ and should not be completely ignored.

It is further to be noted that the method as outlined here is impractical for large numbers for reasons of storage. The optimal $s$ is about $\sqrt{\frac{h}{bK}}$, as we have seen. For factoring numbers of about 256 bits (76 decimal digits) we would expect $h$ to be about $10^{38}$. Even assuming an estimate good to within 1 in one million, and strong congruence conditions on $h$ (such as $h \equiv 0 \pmod{10000}$, for example), the optimal $s$ would be of the order of $10^{14}$. Some compromise would be necessary in any serious implementation, then, trading storage for time.

A small point remains. Having determined $h$, it may yet be possible for us to obtain useless factorings by successive squarings of a form $f^{h'}$. Shanks gives the pathological case of discriminant $-4 \cdot 10721$, with class number 128 and class group $\mathcal{C}(2) \times \mathcal{C}(64)$. There are three ambiguous forms, $(71, 0, 151)$, $(111, 0, 111)$, and $(2, 2, 5361)$. But 125 of the 128 reduced forms are of order 64, and the only 32d power in the group is $(2, 2, 5361)$, which is useless for factoring 10721. However, the group decomposition technique described in Chapter 8 is successful in finding one of the two useful forms. Indeed, given a form $f$ of order 64 (an initial failure to factor), one then has 1 chance in 2 that a second form chosen at random will, with $f$, eventually be determined to generate the entire group and will be reduced to one of the two useful forms.

## 10.4    SPAR

The last of the heuristic class group methods we shall discuss, and the most recently discovered, is the Schnorr-Lenstra [SCHN84] or the SPAR (Shanks-Pollard-Atkin-Rickert) method. As an introduction, we shall first describe the Pollard $p-1$ method [POLL74], from which SPAR in part derives.

### 10.4.1    Pollard $p-1$

Let $N$ be a composite number, and let $p$ denote some (presumably large) prime factor of $N$. The multiplicative group of integers modulo $p$ is cyclic of order $p-1$ and arithmetic done modulo $N$ can be viewed as arithmetic modulo $p$ (except for division by $p$, which can either be considered to be so unlikely as to be impossible, or to be an operation which would generate an error to be trapped and so discovered). If we knew the value of $p-1$, then exponentiating a randomly chosen integer $a$ modulo $N$ would, by Fermat's Little Theorem, produce the factor $p$ of $N$:

If $b \equiv a^{p-1} \pmod{N}$,

then $b \equiv 1 \pmod{N}$

and hence $p$ divides $\gcd(b - 1, N)$.

The problem in this is that knowing the appropriate exponent $p-1$ is equivalent to knowing $p$.

The answer to this problem, in Pollard's method, is to use as an exponent an integer $M$ which is the product of all "small" primes to "large" powers. If the prime factor $p$ of $N$ has the property that $p-1$

contains only small primes, then $p - 1$ will divide $M$, and we will have

$$b \equiv a^M \equiv (a^{p-1})^{M/(p-1)} \equiv 1^{M/(p-1)} \equiv 1 \pmod{N}$$

so that $p$ will appear in the gcd of $b - 1$ and $N$.

The Pollard $p - 1$ technique relies on the following basic properties of the integers modulo $p$:

a) the order of the group is about $p$;

b) by finding the exact order of the group one can find a factor of $N$;

c) computation in the group is not overly difficult.

By stretching the argument for the third property, one can easily see that the same properties apply to the class group of discriminant $-N$ or $-4N$.

A second step of this factoring method is to assume, after having performed the initial exponentiation, that the order of $b$ is a single (large) prime $Q$. If this is the case, then the following method will produce a factor of $N$ by eventually computing $b^Q \pmod{N}$. We first compute $b^{P_1} \pmod{N}$ for the first prime $P_1$ not included in $M$. We compute and save in a list the small even powers of $b$. By multiplying $b^{P_1}$ by the right small even power of $b$ we produce $b^{P_1+2k} = b^{P_2}$ for the second prime $P_2$ not included in $M$. Continuing on, we can produce all prime powers of $b$ in this way with only one multiplication per prime. By multiplying the result into a running product $R$, taken modulo $N$, we can periodically check $\gcd(R - 1, N)$ to determine if we can factor $N$.

## 10.4.2   SPAR

The SPAR class group method is a synthesis of ideas from CLASNO
and from the Pollard method. We first choose a random form $f$ of
discriminant $-N$ or $-4N$. Letting $M$ be the product of all small *odd*
primes to large powers, we then compute $g = f^M$ for this highly composite odd exponent. If it should happen that the odd part of the class
number divides $M$, then $g$ is of order some power of 2, and successively
squaring $g$ will eventually produce an ambiguous form. If we are not so
fortunate, but have instead that $M$ contains all the odd factors of the
class number except for one rather large prime, then we may initiate
a random walk through the group, or a giant-step-baby-step approach,
until we discover this large prime. From that we can go on as before to
obtain an ambiguous form.

One fact which SPAR could use to advantage would be the fact
that class groups need not be cyclic (as the multiplicative group modulo primes must be) and therefore one might be helped by having the
maximal exponent in the class group, much smaller than the order of
the group itself. Unfortunately, as we have seen both from statistical
results and from the suggestions of the Cohen-Lenstra heuristics, class
groups exhibit a strong tendency to be cyclic, and even if not cyclic, to
be nearly so.

In a similar vein, SPAR is aided to a very small degree by the fact
that class numbers are more often divisible by small primes than are
random numbers of comparable size. However, since the excess factors
of primes $p$ are of size $\frac{1}{p^2}$, this is only a small help except for the very
small primes.

# 10.5 CFRAC

For over a decade the "workhorse" factoring method was the continued fraction method, usually referred to as CFRAC, whose practicality was first demonstrated by Morrison and Brillhart [MORR71, MORR75], but whose roots go back somewhat farther. Unlike the previously discussed methods, in which the class group structure is vital to factoring, the continued fraction method only tangentially uses binary quadratic forms. We include it here for completeness.

In CFRAC the basic goal is to exploit the algebraic factoring

$$x^2 - y^2 = (x + y)(x - y) = N.$$

Since it is difficult to solve the algebraic equation, we solve it as a congruence. If

$$x^2 - y^2 \equiv 0 \pmod{N} \tag{10.3}$$

then $\gcd(x + y, N)$ and $\gcd(x - y, N)$ should be nontrivial.

We create a solution to the congruence as follows. As a precomputation, we create a "factor base" $S$ of a few thousand small primes with certain desirable properties. We then collect a series of pairs $(X_i, Y_i)$ such that $X_i^2 \equiv Y_i \pmod{N}$ for each $i$, and for which the $Y_i$ can be completely factored using only primes from the factor base. We then represent the factorings of the $Y_i$ as a matrix of 0's and 1's, with one column for each prime in the factor base, one row for each $Y_i$, and entries 0 or 1 according as the prime divides the corresponding $Y_i$ to an even or odd power. A linear dependence on rows $i_j$, $j = 1, \ldots, k$ of this matrix, which can be found by elimination, now corresponds to the fact that the product

$$\prod_{j=1}^{k} Y_{i_j}$$

is a perfect square $K^2$. We thus have

$$\prod_{j=1}^{k} X_{i_j}^2 \equiv \prod_{j=1}^{k} Y_{i_j} \equiv K^2 \pmod{N}$$

and the algebraic factoring (10.3) of the congruence should yield a factor of $N$.

The expansion of cycles of forms of positive discriminant $4N$, equivalent to expanding the continued fraction for $\sqrt{N}$, is one way to obtain integers $Y_i$, namely the end coefficients of the forms, which are small enough (less than $2\sqrt{N}$) to be reasonably likely to be factorable using the small primes of the factor base, and for which we can readily compute integers $X_i$ such that $X_i^2 \equiv Y_i \pmod{N}$ for each $i$ (see Theorem 3.16).

Using this technique, the property that determines whether or not a small prime $p$ is to be included in the factor base or not is whether or not the residue symbol $\left(\frac{N}{p}\right)$ is $+1$ or not, since only the primes for which $N$ is a residue can divide the end coefficients of forms.

We illustrate this with the example from Morrison and Brillhart, taken from earlier work of Lehmer and Powers [LEHM31], $N = 13290059$. Beginning with $(1, 7290, -4034)$, we cycle forward through the principal cycle of discriminant $4N$. The 6th form is $(-2050, 5794, 2389)$, the 23rd form is $(4633, 6998, -226)$, and the 24th form is $(-226, 7014, 4385)$. The product $(-2050) \cdot 4633 \cdot (-226)$ is $46330^2$, and having carried along the convergents (we only need to compute these modulo $N$), we have that

$$171341^2 + 2050 \equiv 5235158^2 - 4633 \equiv 1914221^2 + 226 \equiv 0 \pmod{13290059}.$$

We find that $171341 \cdot 5235158 \cdot 1914221 \equiv 1496504$, and that $\gcd(1496504+$

$46330, 13290059) = 3119$ and $\gcd(1496504 - 46330, 13290059) = 4261$. These are the two prime factors of 13290059.

# 10.6    A General Analysis

We shall define an integer $x$ to be *y-smooth* if all prime factors of $x$ are less than or equal to $y$, and loosely define $x$ to be *smooth* if $x$ is $y$-smooth for an appropriate $y$ such that a given heuristic method works. The details of estimations of smoothness is complicated [CANF83], and the analysis of the running times of SPAR and CFRAC has appeared in various places, including [POME83], [SCHN85], and [WAGS], so we shall not include the details here. Clearly, though, it is the smoothness of class numbers that affects the running time of SPAR, and the smoothness of the lead coefficients of forms in cycles that affects the running time of CFRAC. To apply the analysis of [CANF83] it is also necessary to make the (we hope reasonable) assumption that these integers are as smooth as random integers of comparable magnitude. As we have seen, class numbers are slightly more smooth, but not so much so as to affect first-order estimates.

For large integers $N$, the probability that $N$ is $\alpha$-smooth is approximately $\alpha^{\frac{1}{\alpha}}$, and the function which measures running times of these algorithms, analyzed according to the smoothness of the relevant integers, is

$$L(N) = \exp(\sqrt{\ln N \ln \ln N})$$

With this definition, it has been shown that, under the assumptions on smoothness of relevant versus random integers, CFRAC takes about $L(N)^{\alpha + o(1)}$ steps to factor $N$, with $\alpha$ a constant between 1 and $\sqrt{3/2}$ [WAGS]. SPAR, on the other hand, has a running time of $L(N)^{1 + o(1)}$.

# Bibliography

[BAKE66]     Alan Baker, "Linear forms in the logarithms of algebraic numbers," *Mathematika*, v. 13, 1966, 204-216. (Chapter 5)

[BRIL83]     John Brillhart, D. H. Lehmer, J. L. Selfridge, B. Tuckerman, and S. S. Wagstaff, Jr., *Factorizations of $b^n \pm 1, b = 2, 3, 5, 6, 7, 10, 11, 12$, up to High Powers*, American Mathematical Society, Providence, R. I., 1983. (Chapter 10)

[BROW71a]    Ezra Brown, "Representation of discriminantal divisors by binary quadratic forms," *Journal of Number Theory*, v. 3, 1971, 213-225. (Chapter 9)

[BROW71b]    Ezra Brown, "A theorem on biquadratic reciprocity," *Proceedings of the American Mathematical Society*, v. 30, 1971, 220-222. (Chapter 9)

[BROW72a]    Ezra Brown, "Binary quadratic forms of determinant $-pq$," *Journal of Number Theory*, v. 4, 1972, 408-410. (Chapter 9)

[BROW72b]    Ezra Brown, "Quadratic forms and biquadratic reciprocity," *Crelle*, v. 253, 1972, 214-220. (Chapter 9)

[BUEL76]     Duncan A. Buell, "Class groups of quadratic fields," *Mathematics of Computation*, v. 30, 1976, 610-623. (Chapters 5, 8)

[BUEL77]     Duncan A. Buell, "Small class numbers and extreme values of $L$-functions of quadratic fields," *Mathematics of Computation*, v. 31, 1977, 786-796. (Chapter 5)

[BUEL78]     Duncan A. Buell, "Computer computation of class groups of quadratic fields," *Congressus Numerantium*, v. 22, 1978, 3-12.

[BUEL84]     Duncan A. Buell, "The expectation of success using a Monte Carlo factoring method–some statistics on quadratic class numbers," *Mathematics of Computation*, v. 43, 1984, 313-327. (Chapters 5, 8)

[BUEL87a]     Duncan A. Buell, "Class groups of quadratic fields II," *Mathematics of Computation*, v. 48, 1987, 85-93. (Chapters 5, 8)

[BUEL87b]     Duncan A. Buell, "Factoring: Algorithms, computers, and computations," *The Journal of Supercomputing*, v. 1, 1987, 191-216. (Chapter 10)

[BURD69]     Klaus Burde, "Ein rationales biquadratisches Reziprozitätsgesetz," *Crelle*, v. 235, 1969, 175-184. (Chapter 9)

[CANF83]     E. R. Canfield, P. Erdös, and C. Pomerance, "On a problem of Oppenheim concerning 'Factorisatio Numerorum'," *Journal of Number Theory*, v. 17, 1983, 1-28. (Chapter 10)

[CANT]     G. Cantor, "Zwei Sätze aus der Theorie der binären quadratischer Formen'," *Z. Math. Physik*, v. 13, 1868, 259-261. (Chapter 9)

[CHOW34]   S. Chowla, "An extension of Heilbronn's class number theorem," *Quart. J. Math. Oxford Ser.*, v. 5, 1934, 304-307. (Chapter 8)

[CHOW74]   S. Chowla and P. Hartung, "Congruence properties of class numbers of quadratic fields," *Journal of Number Theory*, v. 6, 1974, 136-137. (Chapter 8)

[COHE84]   H. Cohen and H. W. Lenstra, Jr., "Heuristics on class groups of number fields," *Number Theory, Lecture Notes in Mathematics #1068*, H. Jager, ed., Berlin, Springer-Verlag, 1984, 33-62. (Chapter 8)

[CRAI73]   Maurice Craig, "A type of class group for imaginary quadratic fields," *Acta Arithmetica*, v. 22, 1973, 449-459. (Chapter 8)

[DIAZ74]   Francisco Diaz y Diaz, "Sur les corps quadratiques dont le 3-rang du groupe des classes est supérieur a 1," *Seminaire Delange-Pisot-Poitou*, G15, 1973/74. (Chapter 8)

[DIAZ78]   Francisco Diaz y Diaz, "Sur le 3-rang des corps quadratiques," Thesis 3rd cycle, Orsay, 1978. (Chapter 8)

[DIAZ79] F. Diaz y Diaz, D. Shanks, and H. C. Williams, "Quadratic fields with 3-rank equal to 4," *Mathematics of Computation*, v. 33, 1979, 836-840. (Chapter 8)

[DICK] L. E. Dickson, *History of the Theory of Numbers*, Chelsea, New York. (Chapter 10)

[DIRI] P. G. L. Dirichlet, "Einige neue Sätze über unbestimmte Gleichungen," *Gesammelte Werke*, Chelsea, New York, 219-236. (Chapter 9)

[GAUS] C. F. Gauss, *Disquisitiones Arithmeticae*, Springer, New York, 1985.

[GERT86] Frank Gerth III, "Limit probabilities for coranks of matrices over $GF(q)$," *Linear and Multilinear Algebra*, v. 19, 1986, 79-93. (Chapter 8)

[HART74] P. Hartung, "Proof of the existence of infinitely many imaginary quadratic fields whose class number is not divisible by 3," *Journal of Number Theory*, v. 6, 1974, 276-278. (Chapter 8)

[HECK] Erich Hecke, *Vorlesungen über die Theorie der Algebraische Zahlen*, Chelsea, New York, 210-217. (Chapter 6)

[HEEG] Kurt Heegner, "Diophantische Analysis und Modulfunktionen," *Mathematische Zeitschrift*, v. 56, 1952, 227-253. (Chapter 5)

[HEIL34]     H. Heilbronn and E. H. Linfoot, "On the imaginary quadratic corpora of class number one," *Quart. J. Math. Oxford Ser.*, v. 5, 1934, 293-301. (Chapter 5)

[HOND68]     Taira Honda, "On real quadratic fields whose class numbers are multiples of 3," *Crelle*, v. 233, 1968, 101-102. (Chapter 8)

[INCE]       E. L. Ince, *Cycles of reduced ideals in quadratic fields*, British Association for the Advancement of Science, London, 1934.

[LEHM31]     D. H. Lehmer and R. E. Powers, "On factoring large numbers," *Bulletin of the American Mathematical Society*, v. 37, 1931, 770-776. (Chapter 10)

[LEHM69]     D. H. Lehmer, "Computer technology applied to the theory of numbers," *MAA Studies in Mathematics, Studies in Number Theory*, W. J. Leveque, ed., v. 6, 1969, 117-151. (Chapter 5)

[LENS]       H. W. Lenstra, Jr., "Factoring integers with elliptic curves," to appear. (Chapter 10)

[LITT28]     J. E. Littlewood, "On the class number of the corpus $P(\sqrt{-k})$, *Proc. London Math. Soc.*, v. 28, 1928, 358-372. (Chapter 5)

[MATH]       G. B. Mathews, *Theory of Numbers*, Chelsea, New York.

[MORR71]    Michael A. Morrison and John Brillhart, "The factor-
            ization of $F_7$," *Bulletin of the American Mathematical
            Society*, v. 77, 1971, 264. (Chapter 10)

[MORR75]    Michael A. Morrison and John Brillhart, "A method of
            factoring and the factorization of $F_7$," *Mathematics of
            Computation*, v. 29, 1975, 183-205. (Chapter 10)

[NAGE22]    T.      Nagell,      "Über      die      Klassenzahl
            imaginär-quadratischer Körper," *Abh. Math. Seminar
            Univ. Hamburg*, v. 1, 1922, 140-150. (Chapter 8)

[NEIL74]    Carol Neild and Daniel Shanks, "On the 3-rank of
            quadratic fields and the Euler product," *Mathematics
            of Computation*, v. 28, 1974, 279-291. (Chapter 8)

[PALL69]    Gordon Pall, "Discriminantal divisors of binary
            quadratic forms," '*Journal of Number Theory*, v. 1,
            1969, 525-533. (Chapter 9)

[POLL74]    J. M. Pollard, "Theorems on factoring and primality
            testing," *Proceedings of the Cambridge Philosophical
            Society*, v. 76, 1974, 521-528. (Chapter 10)

[POME83]    Carl Pomerance, "Analysis and comparison of some in-
            teger factoring algorithms," *Computational Methods in
            Number Theory*, H. W. Lenstra, Jr., R. Tijdeman, eds.,
            Math. Centrum, Amsterdam, 1983, 89-139. (Chapter
            10)

[RIES85]    Hans Riesel, *Prime Numbers and Computer Methods for Factorization*, Birkhäuser, Boston, 1985. (Chapter 10)

[SCHN84]    C. P. Schnorr and H. W. Lenstra, Jr., "A Monte Carlo factoring algorithm with linear storage," *Mathematics of Computation*, v. 43, 1984, 289-312. (Chapter 10)

[SCHO32]    Arnold Scholz, "Über die Beziehung der Klassenzahlen quadratischer Körper zueinander," *Crelle*, v. 166, 1932, 201-203. (Chapter 8)

[SCHO35]    Arnold Scholz, "Über die Lösbarkeit der Gleichung $t^2 - Du^2 = -4$," *Mathematische Zeitschrift*, v. 39, 1935, 95-111. (Chapter 8)

[SCHO83]    R. J. Schoof, "Class groups of complex quadratic fields," *Mathematics of Computation*, v. 41, 1983, 295-302. (Chapter 8)

[SHAN69]    Daniel Shanks, "Class number, a theory of factorization, and genera," *Proc. Symp. in Pure Maths.*, American Mathematical Society, Providence, R. I., v. 20, 1969, 415-440. (Chapters 4, 8, 10)

[SHAN72a]    Daniel Shanks, "New types of quadratic fields having three invariants divisible by three," *Journal of Number Theory*, v. 4, 1972, 537-556. (Chapter 8)

[SHAN72b]    Daniel Shanks and Peter Weinberger, "A quadratic field of prime discriminant requiring three generators for its class group, and related theory," *Acta Arithmetica*, v. 21, 1972, 71-87. (Chapter 8)

[SHAN73a]    Daniel Shanks, "Systematic examination of Little-wood's bounds on $L(1, \chi)$," *Proc. Symp. in Pure Maths.*, American Mathematical Society, Providence, R. I., v. 24, 1973, 267-283. (Chapter 5)

[SHAN73b]    Daniel Shanks and Richard Serafin, "Quadratic fields with four invariants divisible by three," *Mathematics of Computation*, v. 27, 1973, 183-187. (Chapter 8)

[SMIT]       H. J. S. Smith, *Report on the Theory of Numbers*, Chelsea, New York. (Chapter 7)

[SOLD77]     James Solderitsch, "Imaginary quadratic number fields with special class groups," Ph. D. dissertation, Lehigh University, 1977.

[STAR67]     H. M. Stark, "There is no tenth complex quadratic field with class number one," *Michigan Mathematics Journal*, v. 14, 1967, 1-27. (Chapter 5)

[STAR69]     H. M. Stark, "On the 'gap' in a theorem of Heegner," *Journal of Number Theory*, v. 1, 1969, 16-27. (Chapter 5)

[WAGS]       Samuel S. Wagstaff, Jr., "Methods of factoring large integers," to appear.

[WASH86]     Lawrence C. Washington, "Some remarks on Cohen-Lenstra heuristics," *Mathematics of Computation*, v. 47, 1986, 741-747. (Chapter 8)

[WEIN73]     P. J. Weinberger, "Real quadratic fields with class num-
             bers divisible by $n$," *Journal of Number Theory*, v. 5,
             1973, 237-241. (Chapter 8)

[WILL85]     H. C. Williams, "Continued fractions and number-
             theoretic computations," *Rocky Mountain Journal of
             Mathematics*, v. 15, 1985, 621-655. (Chapter 10)

[YAMA70]     Yoshihiko Yamamoto, "On unramified Galois exten-
             sions of quadratic number fields," *Osaka Journal of
             Mathematics*, v. 7, 1970, 57-76. (Chapter 8)

# Appendix 1: Tables, Negative Discriminants

The following two tables are of class numbers of forms for fundamental negative discriminants $-D$ in the range $0 < D < 10000$. For each discriminant we give the value of $D$, the number of classes of forms per genus $H/G$, and the number of genera $G$. The class number is thus $G \cdot H/G$. Even and odd discriminants are presented separately.

*APPENDIX 1*

Table 1A

| Disc | H/G | G | Disc | H/G | G | Disc | H/G | G | Disc | H/G | G | Disc | H/G | G | Disc | H/G | G |
|---|---|---|---|---|---|---|---|---|---|---|---|---|---|---|---|---|---|
| 3 | 1 | 1 | 7 | 1 | 1 | 11 | 1 | 1 | 15 | 1 | 2 | 19 | 1 | 1 | 23 | 3 | 1 |
| 31 | 3 | 1 | 35 | 1 | 2 | 39 | 2 | 2 | 43 | 1 | 1 | 47 | 5 | 1 | 51 | 1 | 2 |
| 55 | 2 | 2 | 59 | 3 | 1 | 67 | 1 | 1 | 71 | 7 | 1 | 79 | 5 | 1 | 83 | 3 | 1 |
| 87 | 3 | 2 | 91 | 1 | 2 | 95 | 4 | 2 | 103 | 5 | 1 | 107 | 3 | 1 | 111 | 4 | 2 |
| 115 | 1 | 2 | 119 | 5 | 2 | 123 | 1 | 2 | 127 | 5 | 1 | 131 | 5 | 1 | 139 | 3 | 1 |
| 143 | 5 | 2 | 151 | 7 | 1 | 155 | 2 | 2 | 159 | 5 | 2 | 163 | 1 | 1 | 167 | 11 | 1 |
| 179 | 5 | 1 | 183 | 4 | 2 | 187 | 1 | 2 | 191 | 13 | 1 | 195 | 1 | 4 | 199 | 9 | 1 |
| 203 | 2 | 2 | 211 | 3 | 1 | 215 | 7 | 2 | 219 | 2 | 2 | 223 | 7 | 1 | 227 | 5 | 1 |
| 231 | 3 | 4 | 235 | 1 | 2 | 239 | 15 | 1 | 247 | 3 | 2 | 251 | 7 | 1 | 255 | 3 | 4 |
| 259 | 2 | 2 | 263 | 13 | 1 | 267 | 1 | 2 | 271 | 11 | 1 | 283 | 3 | 1 | 287 | 7 | 2 |
| 291 | 2 | 2 | 295 | 4 | 2 | 299 | 4 | 2 | 303 | 5 | 2 | 307 | 3 | 1 | 311 | 19 | 1 |
| 319 | 5 | 2 | 323 | 2 | 2 | 327 | 6 | 2 | 331 | 3 | 1 | 335 | 9 | 2 | 339 | 3 | 2 |
| 347 | 5 | 1 | 355 | 2 | 2 | 359 | 19 | 1 | 367 | 9 | 1 | 371 | 4 | 2 | 379 | 3 | 1 |
| 383 | 17 | 1 | 391 | 7 | 2 | 395 | 4 | 2 | 399 | 4 | 4 | 403 | 1 | 2 | 407 | 8. | 2 |
| 411 | 3 | 2 | 415 | 5 | 2 | 419 | 9 | 1 | 427 | 1 | 2 | 431 | 21 | 1 | 435 | 1 | 4 |
| 439 | 15 | 1 | 443 | 5 | 1 | 447 | 7 | 2 | 451 | 3 | 2 | 455 | 5 | 4 | 463 | 7 | 1 |
| 467 | 7 | 1 | 471 | 8 | 2 | 479 | 25 | 1 | 483 | 1 | 4 | 487 | 7 | 1 | 491 | 9 | 1 |
| 499 | 3 | 1 | 503 | 21 | 1 | 511 | 7 | 2 | 515 | 3 | 2 | 519 | 9 | 2 | 523 | 5 | 1 |
| 527 | 9 | 2 | 535 | 7 | 2 | 543 | 6 | 2 | 547 | 3 | 1 | 551 | 13 | 2 | 555 | 1 | 4 |
| 559 | 8 | 2 | 563 | 9 | 1 | 571 | 5 | 1 | 579 | 4 | 2 | 583 | 4 | 2 | 587 | 7 | 1 |
| 591 | 11 | 2 | 595 | 1 | 4 | 599 | 25 | 1 | 607 | 13 | 1 | 611 | 5 | 2 | 615 | 5 | 4 |
| 619 | 5 | 1 | 623 | 11 | 2 | 627 | 1 | 4 | 631 | 13 | 1 | 635 | 5 | 2 | 643 | 3 | 1 |
| 647 | 23 | 1 | 651 | 2 | 4 | 655 | 6 | 2 | 659 | 11 | 1 | 663 | 4 | 4 | 667 | 2 | 2 |
| 671 | 15 | 1 | 679 | 9 | 2 | 683 | 5 | 1 | 687 | 6 | 2 | 691 | 5 | 1 | 695 | 12 | 2 |
| 699 | 5 | 2 | 703 | 7 | 2 | 707 | 3 | 2 | 715 | 1 | 4 | 719 | 31 | 1 | 723 | 2 | 2 |
| 727 | 13 | 1 | 731 | 6 | 2 | 739 | 5 | 1 | 743 | 21 | 1 | 751 | 15 | 1 | 755 | 6 | 2 |
| 759 | 6 | 4 | 763 | 2 | 2 | 767 | 11 | 2 | 771 | 3 | 2 | 779 | 5 | 2 | 787 | 5 | 1 |
| 791 | 16 | 2 | 795 | 1 | 4 | 799 | 8 | 2 | 803 | 5 | 2 | 807 | 7 | 2 | 811 | 7 | 1 |
| 815 | 15 | 2 | 823 | 9 | 1 | 827 | 7 | 1 | 831 | 14 | 2 | 835 | 3 | 2 | 839 | 33 | 1 |
| 843 | 3 | 2 | 851 | 5 | 2 | 859 | 7 | 1 | 863 | 21 | 1 | 871 | 11 | 2 | 879 | 11 | 2 |
| 883 | 3 | 1 | 887 | 29 | 1 | 895 | 8 | 2 | 899 | 7 | 2 | 903 | 4 | 4 | 907 | 3 | 1 |
| 911 | 31 | 1 | 915 | 2 | 4 | 919 | 19 | 1 | 923 | 5 | 2 | 935 | 7 | 4 | 939 | 4 | 2 |
| 943 | 8 | 2 | 947 | 5 | 1 | 951 | 13 | 2 | 955 | 2 | 2 | 959 | 18 | 2 | 967 | 11 | 1 |
| 971 | 15 | 1 | 979 | 4 | 2 | 983 | 27 | 1 | 987 | 2 | 4 | 991 | 17 | 1 | 995 | 4 | 2 |
| 1003 | 2 | 2 | 1007 | 15 | 2 | 1011 | 6 | 2 | 1015 | 4 | 4 | 1019 | 13 | 1 | 1023 | 4 | 4 |
| 1027 | 2 | 2 | 1031 | 35 | 1 | 1039 | 23 | 1 | 1043 | 4 | 2 | 1047 | 8 | 2 | 1051 | 5 | 1 |
| 1055 | 18 | 2 | 1059 | 3 | 2 | 1063 | 19 | 1 | 1067 | 6 | 2 | 1079 | 17 | 2 | 1087 | 9 | 1 |
| 1091 | 17 | 1 | 1095 | 7 | 4 | 1099 | 3 | 2 | 1103 | 23 | 1 | 1111 | 11 | 2 | 1115 | 5 | 2 |
| 1119 | 16 | 2 | 1123 | 5 | 1 | 1131 | 2 | 4 | 1135 | 9 | 2 | 1139 | 8 | 2 | 1147 | 3 | 2 |
| 1151 | 41 | 1 | 1155 | 1 | 8 | 1159 | 8 | 2 | 1163 | 7 | 1 | 1167 | 11 | 2 | 1171 | 7 | 1 |
| 1187 | 9 | 1 | 1191 | 12 | 2 | 1195 | 4 | 2 | 1199 | 19 | 2 | 1203 | 3 | 2 | 1207 | 9 | 2 |
| 1211 | 7 | 2 | 1219 | 3 | 2 | 1223 | 35 | 1 | 1227 | 2 | 2 | 1231 | 27 | 1 | 1235 | 3 | 4 |
| 1239 | 8 | 4 | 1243 | 2 | 2 | 1247 | 13 | 2 | 1255 | 6 | 2 | 1259 | 15 | 1 | 1263 | 10 | 2 |
| 1267 | 3 | 2 | 1271 | 20 | 2 | 1279 | 23 | 1 | 1283 | 11 | 2 | 1291 | 9 | 1 | 1295 | 9 | 4 |
| 1299 | 4 | 2 | 1303 | 11 | 1 | 1307 | 11 | 1 | 1311 | 7 | 4 | 1315 | 3 | 2 | 1319 | 45 | 1 |
| 1327 | 15 | 1 | 1335 | 7 | 4 | 1339 | 4 | 2 | 1343 | 17 | 2 | 1347 | 3 | 2 | 1351 | 12 | 2 |
| 1355 | 6 | 2 | 1363 | 3 | 2 | 1367 | 25 | 1 | 1371 | 6 | 2 | 1379 | 8 | 2 | 1383 | 9 | 2 |
| 1387 | 2 | 2 | 1391 | 22 | 2 | 1399 | 27 | 1 | 1403 | 7 | 2 | 1407 | 6 | 4 | 1411 | 2 | 2 |
| 1415 | 17 | 2 | 1419 | 3 | 4 | 1423 | 9 | 2 | 1427 | 15 | 1 | 1435 | 1 | 4 | 1439 | 39 | 1 |
| 1443 | 2 | 4 | 1447 | 23 | 1 | 1451 | 13 | 1 | 1455 | 7 | 4 | 1459 | 11 | 1 | 1463 | 8 | 4 |
| 1471 | 23 | 1 | 1479 | 7 | 4 | 1483 | 7 | 1 | 1487 | 37 | 1 | 1491 | 3 | 4 | 1495 | 5 | 4 |
| 1499 | 13 | 1 | 1507 | 2 | 2 | 1511 | 49 | 1 | 1515 | 3 | 4 | 1523 | 7 | 1 | 1527 | 7 | 2 |
| 1531 | 11 | 1 | 1535 | 19 | 2 | 1543 | 19 | 1 | 1547 | 3 | 4 | 1551 | 8 | 4 | 1555 | 2 | 2 |
| 1559 | 51 | 1 | 1563 | 3 | 2 | 1567 | 15 | 1 | 1571 | 17 | 1 | 1579 | 9 | 1 | 1583 | 33 | 1 |
| 1591 | 11 | 2 | 1595 | 4 | 4 | 1599 | 9 | 4 | 1603 | 3 | 2 | 1607 | 27 | 1 | 1615 | 6 | 4 |

Table 1A

| Disc | H/G | G | Disc | H/G | G | Disc | H/G | G | Disc | H/G | G | Disc | H/G | G | Disc | H/G | G |
|---|---|---|---|---|---|---|---|---|---|---|---|---|---|---|---|---|---|
| 1619 | 15 | 1 | 1623 | 14 | 2 | 1627 | 7 | 1 | 1631 | 22 | 2 | 1635 | 2 | 4 | 1639 | 11 | 2 |
| 1643 | 5 | 2 | 1651 | 4 | 2 | 1655 | 22 | 2 | 1659 | 2 | 4 | 1663 | 17 | 1 | 1667 | 13 | 1 |
| 1671 | 19 | 2 | 1679 | 26 | 2 | 1687 | 9 | 2 | 1691 | 9 | 2 | 1695 | 5 | 4 | 1699 | 11 | 1 |
| 1703 | 14 | 2 | 1707 | 5 | 2 | 1711 | 14 | 2 | 1723 | 5 | 1 | 1727 | 18 | 2 | 1731 | 4 | 2 |
| 1735 | 13 | 2 | 1739 | 10 | 2 | 1743 | 6 | 4 | 1747 | 5 | 1 | 1751 | 24 | 2 | 1759 | 27 | 1 |
| 1763 | 6 | 2 | 1767 | 8 | 4 | 1771 | 2 | 4 | 1779 | 5 | 2 | 1783 | 17 | 1 | 1787 | 7 | 1 |
| 1795 | 4 | 2 | 1799 | 25 | 2 | 1803 | 4 | 2 | 1807 | 6 | 2 | 1811 | 23 | 2 | 1819 | 5 | 2 |
| 1823 | 45 | 1 | 1831 | 19 | 2 | 1835 | 5 | 2 | 1839 | 20 | 2 | 1843 | 3 | 2 | 1847 | 43 | 1 |
| 1851 | 7 | 2 | 1855 | 7 | 4 | 1867 | 5 | 1 | 1871 | 45 | 1 | 1879 | 27 | 1 | 1883 | 7 | 2 |
| 1887 | 5 | 4 | 1891 | 5 | 2 | 1895 | 24 | 2 | 1903 | 11 | 2 | 1907 | 13 | 1 | 1915 | 3 | 2 |
| 1919 | 22 | 2 | 1923 | 5 | 2 | 1927 | 9 | 2 | 1931 | 21 | 1 | 1939 | 4 | 2 | 1943 | 16 | 2 |
| 1947 | 2 | 4 | 1951 | 33 | 1 | 1955 | 3 | 4 | 1959 | 21 | 2 | 1963 | 3 | 2 | 1967 | 18 | 2 |
| 1979 | 23 | 1 | 1983 | 8 | 2 | 1987 | 7 | 1 | 1991 | 28 | 2 | 1995 | 1 | 8 | 1999 | 27 | 1 |
| 2003 | 9 | 1 | 2011 | 7 | 1 | 2015 | 13 | 4 | 2019 | 8 | 2 | 2027 | 11 | 1 | 2031 | 19 | 2 |
| 2035 | 2 | 4 | 2039 | 45 | 1 | 2047 | 9 | 2 | 2051 | 9 | 2 | 2055 | 7 | 4 | 2059 | 4 | 2 |
| 2063 | 45 | 1 | 2067 | 2 | 4 | 2071 | 15 | 2 | 2083 | 7 | 1 | 2087 | 35 | 1 | 2091 | 3 | 4 |
| 2095 | 8 | 2 | 2099 | 19 | 2 | 2103 | 17 | 2 | 2111 | 49 | 1 | 2119 | 17 | 2 | 2123 | 7 | 2 |
| 2127 | 14 | 2 | 2131 | 13 | 1 | 2135 | 11 | 4 | 2139 | 2 | 4 | 2143 | 13 | 1 | 2147 | 7 | 2 |
| 2155 | 6 | 2 | 2159 | 30 | 2 | 2163 | 2 | 4 | 2167 | 9 | 2 | 2171 | 7 | 2 | 2179 | 7 | 1 |
| 2183 | 21 | 2 | 2191 | 15 | 2 | 2195 | 8 | 2 | 2199 | 18 | 2 | 2203 | 5 | 1 | 2207 | 39 | 1 |
| 2211 | 4 | 4 | 2215 | 11 | 2 | 2219 | 12 | 2 | 2227 | 3 | 2 | 2231 | 29 | 2 | 2235 | 3 | 4 |
| 2239 | 35 | 1 | 2243 | 15 | 1 | 2247 | 5 | 4 | 2251 | 7 | 1 | 2255 | 10 | 4 | 2263 | 11 | 2 |
| 2267 | 11 | 1 | 2271 | 22 | 2 | 2279 | 28 | 2 | 2283 | 3 | 2 | 2287 | 29 | 1 | 2291 | 9 | 2 |
| 2307 | 4 | 2 | 2311 | 29 | 1 | 2315 | 9 | 2 | 2319 | 15 | 2 | 2323 | 4 | 2 | 2327 | 24 | 2 |
| 2335 | 7 | 2 | 2339 | 19 | 1 | 2343 | 8 | 4 | 2347 | 5 | 1 | 2351 | 63 | 1 | 2355 | 3 | 4 |
| 2359 | 14 | 2 | 2363 | 5 | 2 | 2371 | 13 | 1 | 2379 | 4 | 4 | 2383 | 29 | 1 | 2387 | 3 | 4 |
| 2391 | 17 | 2 | 2395 | 4 | 2 | 2399 | 59 | 1 | 2407 | 10 | 2 | 2411 | 23 | 1 | 2415 | 5 | 8 |
| 2419 | 4 | 2 | 2423 | 33 | 1 | 2427 | 7 | 2 | 2431 | 7 | 4 | 2435 | 11 | 2 | 2443 | 3 | 2 |
| 2447 | 37 | 1 | 2451 | 2 | 4 | 2455 | 14 | 2 | 2459 | 19 | 1 | 2463 | 17 | 2 | 2467 | 7 | 1 |
| 2471 | 31 | 2 | 2479 | 12 | 2 | 2483 | 10 | 2 | 2487 | 10 | 2 | 2491 | 6 | 2 | 2495 | 28 | 2 |
| 2503 | 21 | 1 | 2507 | 7 | 2 | 2515 | 3 | 2 | 2519 | 32 | 2 | 2531 | 17 | 1 | 2539 | 11 | 1 |
| 2543 | 35 | 1 | 2551 | 41 | 1 | 2555 | 3 | 4 | 2559 | 20 | 2 | 2563 | 3 | 2 | 2567 | 22 | 2 |
| 2571 | 7 | 2 | 2579 | 21 | 1 | 2587 | 4 | 2 | 2591 | 57 | 1 | 2595 | 3 | 4 | 2599 | 15 | 2 |
| 2603 | 10 | 2 | 2607 | 7 | 4 | 2611 | 4 | 2 | 2615 | 23 | 2 | 2623 | 11 | 2 | 2627 | 6 | 2 |
| 2631 | 24 | 2 | 2635 | 3 | 4 | 2639 | 16 | 2 | 2643 | 5 | 2 | 2647 | 15 | 1 | 2651 | 13 | 2 |
| 2659 | 13 | 1 | 2663 | 43 | 1 | 2667 | 2 | 4 | 2671 | 23 | 1 | 2679 | 13 | 4 | 2683 | 5 | 1 |
| 2687 | 51 | 1 | 2699 | 15 | 1 | 2703 | 7 | 4 | 2707 | 7 | 1 | 2711 | 53 | 1 | 2715 | 2 | 4 |
| 2719 | 41 | 1 | 2723 | 6 | 2 | 2731 | 11 | 1 | 2735 | 31 | 2 | 2739 | 4 | 4 | 2743 | 10 | 2 |
| 2747 | 9 | 2 | 2751 | 10 | 4 | 2755 | 2 | 4 | 2759 | 27 | 1 | 2767 | 21 | 2 | 2771 | 13 | 2 |
| 2779 | 7 | 2 | 2787 | 3 | 2 | 2791 | 39 | 1 | 2795 | 3 | 4 | 2803 | 9 | 1 | 2807 | 26 | 2 |
| 2811 | 8 | 2 | 2815 | 11 | 2 | 2819 | 21 | 1 | 2823 | 13 | 2 | 2827 | 4 | 2 | 2831 | 34 | 2 |
| 2839 | 13 | 2 | 2843 | 15 | 1 | 2847 | 8 | 4 | 2851 | 11 | 1 | 2855 | 30 | 2 | 2859 | 9 | 2 |
| 2863 | 11 | 2 | 2867 | 6 | 2 | 2879 | 57 | 1 | 2887 | 25 | 1 | 2895 | 9 | 4 | 2899 | 5 | 2 |
| 2903 | 59 | 1 | 2911 | 21 | 2 | 2915 | 6 | 4 | 2919 | 10 | 4 | 2923 | 3 | 2 | 2927 | 31 | 1 |
| 2931 | 7 | 2 | 2935 | 11 | 2 | 2939 | 29 | 1 | 2947 | 4 | 2 | 2951 | 27 | 2 | 2955 | 3 | 4 |
| 2959 | 20 | 2 | 2963 | 13 | 1 | 2967 | 11 | 4 | 2971 | 11 | 1 | 2983 | 10 | 2 | 2987 | 10 | 2 |
| 2991 | 24 | 2 | 2995 | 4 | 2 | 2999 | 73 | 1 | 3003 | 1 | 8 | 3007 | 10 | 2 | 3011 | 21 | 1 |
| 3019 | 7 | 1 | 3023 | 47 | 1 | 3027 | 6 | 2 | 3031 | 17 | 2 | 3035 | 9 | 2 | 3039 | 21 | 2 |
| 3043 | 6 | 2 | 3047 | 19 | 2 | 3055 | 9 | 4 | 3059 | 6 | 4 | 3063 | 8 | 2 | 3067 | 7 | 1 |
| 3071 | 38 | 2 | 3079 | 41 | 1 | 3083 | 13 | 1 | 3091 | 5 | 2 | 3095 | 24 | 2 | 3099 | 10 | 2 |
| 3103 | 10 | 2 | 3107 | 9 | 2 | 3111 | 13 | 4 | 3115 | 3 | 4 | 3119 | 69 | 1 | 3127 | 12 | 2 |
| 3131 | 10 | 2 | 3135 | 5 | 8 | 3139 | 5 | 2 | 3143 | 28 | 2 | 3147 | 5 | 2 | 3151 | 11 | 2 |
| 3155 | 10 | 2 | 3163 | 9 | 1 | 3167 | 53 | 1 | 3171 | 4 | 4 | 3183 | 17 | 2 | 3187 | 7 | 1 |
| 3191 | 69 | 1 | 3199 | 16 | 2 | 3203 | 11 | 1 | 3207 | 16 | 2 | 3215 | 25 | 2 | 3219 | 5 | 4 |
| 3223 | 15 | 2 | 3227 | 7 | 2 | 3235 | 3 | 2 | 3239 | 35 | 2 | 3243 | 2 | 4 | 3247 | 16 | 2 |

Table 1A

| Disc | H/G | G | Disc | H/G | G | Disc | H/G | G | Disc | H/G | G | Disc | H/G | G | Disc | H/G | G |
|---|---|---|---|---|---|---|---|---|---|---|---|---|---|---|---|---|---|
| 3251 | 31 | 1 | 3255 | 5 | 8 | 3259 | 9 | 1 | 3263 | 24 | 2 | 3271 | 27 | 1 | 3279 | 26 | 2 |
| 3287 | 17 | 2 | 3291 | 5 | 2 | 3295 | 16 | 2 | 3299 | 27 | 1 | 3307 | 9 | 1 | 3311 | 18 | 4 |
| 3315 | 1 | 8 | 3319 | 41 | 1 | 3323 | 17 | 1 | 3327 | 13 | 2 | 3331 | 15 | 1 | 3335 | 16 | 4 |
| 3343 | 19 | 1 | 3347 | 11 | 1 | 3351 | 30 | 2 | 3355 | 2 | 4 | 3359 | 69 | 1 | 3363 | 4 | 4 |
| 3367 | 5 | 4 | 3371 | 21 | 1 | 3379 | 8 | 2 | 3383 | 23 | 2 | 3387 | 6 | 2 | 3391 | 37 | 1 |
| 3395 | 5 | 4 | 3399 | 10 | 4 | 3403 | 4 | 2 | 3407 | 57 | 1 | 3415 | 19 | 2 | 3419 | 14 | 2 |
| 3423 | 10 | 4 | 3427 | 3 | 2 | 3431 | 33 | 2 | 3435 | 4 | 4 | 3439 | 15 | 2 | 3443 | 8 | 2 |
| 3451 | 3 | 4 | 3455 | 26 | 1 | 3459 | 6 | 2 | 3463 | 19 | 1 | 3467 | 19 | 1 | 3471 | 15 | 4 |
| 3487 | 14 | 2 | 3491 | 23 | 1 | 3495 | 9 | 4 | 3499 | 11 | 1 | 3503 | 26 | 2 | 3507 | 2 | 4 |
| 3511 | 41 | 1 | 3515 | 5 | 4 | 3523 | 3 | 2 | 3527 | 65 | 1 | 3531 | 4 | 4 | 3535 | 7 | 4 |
| 3539 | 23 | 1 | 3543 | 9 | 2 | 3547 | 9 | 1 | 3551 | 29 | 2 | 3559 | 45 | 1 | 3563 | 11 | 2 |
| 3567 | 5 | 4 | 3571 | 15 | 2 | 3579 | 7 | 2 | 3583 | 29 | 1 | 3587 | 11 | 2 | 3595 | 4 | 2 |
| 3599 | 33 | 2 | 3603 | 8 | 2 | 3607 | 19 | 1 | 3611 | 13 | 2 | 3615 | 12 | 4 | 3619 | 3 | 4 |
| 3623 | 45 | 1 | 3631 | 43 | 1 | 3635 | 5 | 2 | 3639 | 30 | 2 | 3643 | 9 | 1 | 3647 | 27 | 2 |
| 3651 | 9 | 2 | 3655 | 5 | 4 | 3659 | 29 | 1 | 3667 | 5 | 2 | 3671 | 81 | 1 | 3679 | 16 | 2 |
| 3683 | 5 | 2 | 3687 | 21 | 1 | 3691 | 13 | 1 | 3695 | 36 | 2 | 3707 | 7 | 2 | 3711 | 18 | 2 |
| 3715 | 7 | 2 | 3719 | 67 | 1 | 3723 | 3 | 4 | 3727 | 31 | 1 | 3731 | 7 | 4 | 3739 | 11 | 1 |
| 3743 | 28 | 2 | 3747 | 6 | 2 | 3755 | 10 | 2 | 3759 | 13 | 4 | 3763 | 3 | 2 | 3767 | 39 | 1 |
| 3779 | 31 | 1 | 3783 | 12 | 4 | 3787 | 4 | 2 | 3791 | 34 | 2 | 3795 | 2 | 8 | 3799 | 23 | 2 |
| 3803 | 15 | 1 | 3811 | 5 | 2 | 3815 | 14 | 4 | 3819 | 4 | 4 | 3823 | 29 | 1 | 3827 | 11 | 2 |
| 3831 | 21 | 2 | 3835 | 3 | 4 | 3839 | 41 | 2 | 3847 | 23 | 1 | 3851 | 25 | 1 | 3855 | 11 | 4 |
| 3859 | 5 | 2 | 3863 | 61 | 1 | 3867 | 7 | 2 | 3883 | 4 | 2 | 3891 | 12 | 2 | 3895 | 8 | 4 |
| 3899 | 12 | 2 | 3903 | 13 | 2 | 3907 | 7 | 1 | 3911 | 83 | 1 | 3919 | 39 | 1 | 3923 | 23 | 1 |
| 3927 | 5 | 8 | 3931 | 11 | 1 | 3935 | 33 | 2 | 3939 | 4 | 4 | 3943 | 27 | 1 | 3947 | 17 | 1 |
| 3955 | 3 | 4 | 3959 | 34 | 2 | 3963 | 4 | 2 | 3967 | 33 | 1 | 3979 | 10 | 2 | 3983 | 22 | 2 |
| 3991 | 15 | 2 | 3995 | 7 | 4 | 3999 | 12 | 4 | 4003 | 13 | 1 | 4007 | 57 | 1 | 4011 | 5 | 4 |
| 4015 | 5 | 4 | 4019 | 19 | 1 | 4027 | 9 | 1 | 4031 | 42 | 2 | 4035 | 3 | 4 | 4039 | 21 | 2 |
| 4043 | 8 | 2 | 4047 | 10 | 4 | 4051 | 11 | 1 | 4055 | 34 | 2 | 4063 | 12 | 2 | 4071 | 10 | 4 |
| 4079 | 85 | 1 | 4083 | 5 | 2 | 4087 | 15 | 2 | 4091 | 33 | 1 | 4099 | 15 | 1 | 4103 | 21 | 2 |
| 4111 | 39 | 1 | 4115 | 11 | 2 | 4119 | 27 | 2 | 4123 | 2 | 4 | 4127 | 49 | 1 | 4135 | 23 | 2 |
| 4139 | 19 | 1 | 4143 | 22 | 2 | 4147 | 3 | 4 | 4151 | 37 | 2 | 4155 | 3 | 4 | 4159 | 31 | 1 |
| 4163 | 11 | 2 | 4171 | 8 | 2 | 4179 | 4 | 4 | 4183 | 14 | 2 | 4187 | 7 | 2 | 4191 | 15 | 4 |
| 4195 | 4 | 2 | 4199 | 22 | 4 | 4207 | 13 | 2 | 4211 | 23 | 1 | 4215 | 11 | 4 | 4219 | 15 | 1 |
| 4223 | 22 | 2 | 4227 | 5 | 2 | 4231 | 51 | 1 | 4243 | 9 | 1 | 4247 | 31 | 2 | 4251 | 4 | 4 |
| 4255 | 9 | 4 | 4259 | 35 | 1 | 4267 | 4 | 2 | 4271 | 65 | 1 | 4279 | 14 | 2 | 4283 | 21 | 1 |
| 4287 | 24 | 2 | 4291 | 6 | 2 | 4295 | 24 | 2 | 4299 | 9 | 2 | 4303 | 17 | 2 | 4307 | 9 | 2 |
| 4315 | 7 | 2 | 4319 | 42 | 2 | 4323 | 2 | 4 | 4327 | 19 | 1 | 4331 | 17 | 2 | 4339 | 17 | 1 |
| 4343 | 32 | 2 | 4351 | 22 | 2 | 4355 | 5 | 4 | 4359 | 30 | 2 | 4363 | 9 | 1 | 4367 | 34 | 2 |
| 4371 | 5 | 4 | 4379 | 8 | 2 | 4387 | 4 | 2 | 4391 | 79 | 1 | 4395 | 5 | 4 | 4399 | 25 | 2 |
| 4403 | 5 | 4 | 4407 | 10 | 4 | 4411 | 6 | 2 | 4415 | 33 | 2 | 4423 | 33 | 1 | 4427 | 8 | 2 |
| 4431 | 12 | 4 | 4435 | 5 | 2 | 4439 | 38 | 2 | 4443 | 7 | 2 | 4447 | 17 | 1 | 4451 | 29 | 1 |
| 4463 | 55 | 1 | 4467 | 6 | 2 | 4471 | 22 | 2 | 4479 | 35 | 2 | 4483 | 9 | 1 | 4487 | 30 | 2 |
| 4495 | 8 | 4 | 4499 | 17 | 2 | 4503 | 6 | 4 | 4507 | 13 | 1 | 4511 | 42 | 2 | 4515 | 2 | 8 |
| 4519 | 29 | 1 | 4523 | 21 | 1 | 4531 | 6 | 4 | 4535 | 35 | 2 | 4539 | 5 | 4 | 4543 | 7 | 4 |
| 4547 | 17 | 1 | 4551 | 14 | 4 | 4555 | 6 | 2 | 4559 | 36 | 2 | 4567 | 33 | 1 | 4571 | 14 | 2 |
| 4579 | 5 | 2 | 4583 | 61 | 1 | 4587 | 3 | 4 | 4591 | 49 | 1 | 4595 | 12 | 2 | 4603 | 7 | 1 |
| 4607 | 20 | 2 | 4611 | 7 | 4 | 4615 | 9 | 4 | 4619 | 18 | 2 | 4623 | 9 | 4 | 4627 | 5 | 2 |
| 4631 | 38 | 2 | 4639 | 51 | 1 | 4643 | 13 | 1 | 4647 | 13 | 1 | 4651 | 17 | 1 | 4659 | 7 | 2 |
| 4663 | 33 | 1 | 4667 | 11 | 2 | 4679 | 91 | 1 | 4683 | 4 | 4 | 4687 | 8 | 2 | 4691 | 21 | 1 |
| 4695 | 11 | 4 | 4699 | 6 | 2 | 4703 | 75 | 1 | 4711 | 22 | 2 | 4715 | 6 | 4 | 4723 | 9 | 1 |
| 4727 | 27 | 2 | 4731 | 5 | 4 | 4735 | 21 | 2 | 4739 | 13 | 2 | 4747 | 4 | 2 | 4751 | 91 | 1 |
| 4755 | 3 | 4 | 4759 | 55 | 1 | 4763 | 10 | 2 | 4767 | 16 | 2 | 4771 | 6 | 2 | 4783 | 23 | 1 |
| 4787 | 25 | 1 | 4791 | 34 | 2 | 4795 | 3 | 4 | 4799 | 63 | 1 | 4803 | 7 | 2 | 4807 | 10 | 4 |
| 4811 | 11 | 2 | 4819 | 9 | 2 | 4823 | 12 | 4 | 4827 | 6 | 2 | 4831 | 33 | 1 | 4835 | 15 | 4 |
| 4839 | 29 | 2 | 4843 | 4 | 2 | 4847 | 37 | 2 | 4855 | 10 | 2 | 4859 | 15 | 2 | 4863 | 26 | 2 |

Table 1A

| Disc | H/G | G | Disc | H/G | G | Disc | H/G | G | Disc | H/G | G | Disc | H/G | G | Disc | H/G | G |
|---|---|---|---|---|---|---|---|---|---|---|---|---|---|---|---|---|---|
| 4867 | 4 | 2 | 4871 | 91 | 1 | 4879 | 13 | 4 | 4883 | 9 | 2 | 4891 | 10 | 2 | 4895 | 16 | 4 |
| 4899 | 4 | 4 | 4903 | 27 | 1 | 4907 | 6 | 2 | 4911 | 25 | 2 | 4915 | 5 | 2 | 4919 | 91 | 1 |
| 4927 | 17 | 2 | 4931 | 35 | 1 | 4935 | 6 | 8 | 4939 | 8 | 2 | 4943 | 55 | 1 | 4947 | 3 | 4 |
| 4951 | 31 | 1 | 4955 | 14 | 2 | 4963 | 6 | 2 | 4967 | 59 | 1 | 4971 | 8 | 2 | 4979 | 15 | 2 |
| 4983 | 13 | 4 | 4987 | 9 | 1 | 4991 | 23 | 4 | 4999 | 33 | 1 | 5003 | 15 | 1 | 5007 | 18 | 2 |
| 5011 | 21 | 1 | 5015 | 22 | 4 | 5019 | 5 | 4 | 5023 | 25 | 1 | 5027 | 7 | 2 | 5035 | 3 | 4 |
| 5039 | 83 | 1 | 5051 | 29 | 1 | 5055 | 9 | 4 | 5059 | 19 | 1 | 5063 | 18 | 2 | 5071 | 25 | 2 |
| 5079 | 32 | 2 | 5083 | 2 | 4 | 5087 | 69 | 1 | 5091 | 7 | 2 | 5095 | 24 | 2 | 5099 | 39 | 1 |
| 5107 | 7 | 1 | 5111 | 39 | 2 | 5115 | 2 | 8 | 5119 | 39 | 1 | 5123 | 13 | 2 | 5127 | 19 | 2 |
| 5131 | 5 | 2 | 5135 | 14 | 4 | 5143 | 20 | 2 | 5147 | 19 | 1 | 5151 | 17 | 4 | 5155 | 6 | 2 |
| 5159 | 22 | 4 | 5163 | 5 | 2 | 5167 | 33 | 1 | 5171 | 35 | 1 | 5179 | 11 | 1 | 5183 | 40 | 2 |
| 5187 | 2 | 8 | 5191 | 20 | 2 | 5195 | 8 | 2 | 5199 | 37 | 2 | 5207 | 28 | 2 | 5215 | 8 | 4 |
| 5219 | 12 | 2 | 5223 | 16 | 2 | 5227 | 15 | 1 | 5231 | 75 | 1 | 5235 | 6 | 4 | 5251 | 7 | 2 |
| 5255 | 46 | 2 | 5259 | 6 | 2 | 5263 | 14 | 2 | 5267 | 7 | 2 | 5271 | 16 | 4 | 5279 | 87 | 1 |
| 5287 | 17 | 2 | 5291 | 9 | 4 | 5295 | 15 | 4 | 5299 | 6 | 2 | 5303 | 55 | 1 | 5307 | 3 | 4 |
| 5311 | 23 | 2 | 5315 | 9 | 2 | 5323 | 15 | 1 | 5327 | 29 | 1 | 5331 | 10 | 2 | 5335 | 8 | 4 |
| 5339 | 17 | 2 | 5343 | 9 | 4 | 5347 | 13 | 1 | 5351 | 93 | 1 | 5359 | 24 | 2 | 5363 | 8 | 2 |
| 5367 | 20 | 2 | 5371 | 6 | 2 | 5379 | 6 | 4 | 5383 | 11 | 2 | 5387 | 23 | 1 | 5395 | 3 | 4 |
| 5399 | 79 | 1 | 5403 | 8 | 2 | 5407 | 43 | 1 | 5411 | 17 | 2 | 5419 | 13 | 1 | 5423 | 17 | 4 |
| 5431 | 57 | 1 | 5435 | 13 | 2 | 5443 | 9 | 1 | 5447 | 30 | 1 | 5451 | 7 | 2 | 5455 | 18 | 2 |
| 5459 | 17 | 2 | 5467 | 2 | 4 | 5471 | 71 | 1 | 5479 | 43 | 1 | 5483 | 17 | 1 | 5487 | 13 | 4 |
| 5495 | 21 | 4 | 5503 | 25 | 1 | 5507 | 23 | 1 | 5511 | 14 | 4 | 5515 | 5 | 2 | 5519 | 97 | 1 |
| 5523 | 3 | 4 | 5527 | 19 | 1 | 5531 | 23 | 1 | 5539 | 9 | 2 | 5543 | 39 | 2 | 5551 | 13 | 4 |
| 5555 | 6 | 4 | 5559 | 11 | 4 | 5563 | 15 | 1 | 5567 | 27 | 2 | 5579 | 15 | 2 | 5583 | 16 | 2 |
| 5587 | 4 | 2 | 5591 | 99 | 1 | 5595 | 3 | 4 | 5599 | 28 | 2 | 5603 | 11 | 2 | 5611 | 5 | 2 |
| 5615 | 31 | 4 | 5619 | 14 | 2 | 5623 | 33 | 1 | 5627 | 14 | 2 | 5631 | 37 | 2 | 5639 | 87 | 1 |
| 5647 | 21 | 1 | 5651 | 31 | 1 | 5655 | 7 | 8 | 5659 | 19 | 1 | 5663 | 28 | 2 | 5667 | 5 | 2 |
| 5671 | 22 | 2 | 5683 | 11 | 1 | 5691 | 6 | 4 | 5695 | 11 | 4 | 5699 | 12 | 2 | 5703 | 27 | 2 |
| 5707 | 4 | 2 | 5711 | 109 | 1 | 5719 | 14 | 4 | 5723 | 7 | 2 | 5727 | 10 | 4 | 5731 | 10 | 2 |
| 5735 | 13 | 4 | 5739 | 11 | 2 | 5743 | 29 | 1 | 5747 | 12 | 2 | 5755 | 6 | 2 | 5759 | 54 | 2 |
| 5763 | 3 | 4 | 5767 | 18 | 2 | 5771 | 22 | 2 | 5779 | 13 | 1 | 5783 | 53 | 1 | 5791 | 33 | 1 |
| 5795 | 8 | 4 | 5799 | 40 | 2 | 5803 | 5 | 2 | 5807 | 65 | 1 | 5811 | 3 | 4 | 5815 | 25 | 2 |
| 5827 | 15 | 1 | 5835 | 3 | 4 | 5839 | 37 | 1 | 5843 | 25 | 1 | 5847 | 25 | 2 | 5851 | 21 | 1 |
| 5855 | 38 | 2 | 5863 | 7 | 4 | 5867 | 21 | 1 | 5871 | 18 | 4 | 5879 | 101 | 1 | 5883 | 4 | 4 |
| 5891 | 13 | 2 | 5899 | 9 | 2 | 5903 | 73 | 1 | 5907 | 4 | 4 | 5911 | 28 | 2 | 5919 | 27 | 2 |
| 5923 | 7 | 1 | 5927 | 71 | 1 | 5935 | 21 | 2 | 5939 | 35 | 1 | 5943 | 8 | 4 | 5947 | 4 | 2 |
| 5951 | 45 | 2 | 5955 | 5 | 4 | 5959 | 21 | 2 | 5963 | 12 | 2 | 5971 | 7 | 2 | 5979 | 12 | 2 |
| 5983 | 22 | 2 | 5987 | 15 | 1 | 5991 | 35 | 2 | 5995 | 4 | 4 | 5999 | 45 | 2 | 6007 | 27 | 1 |
| 6011 | 27 | 1 | 6015 | 9 | 2 | 6019 | 11 | 2 | 6023 | 41 | 2 | 6031 | 25 | 2 | 6035 | 7 | 4 |
| 6043 | 9 | 1 | 6047 | 71 | 1 | 6051 | 10 | 2 | 6055 | 9 | 4 | 6059 | 13 | 2 | 6063 | 9 | 4 |
| 6067 | 15 | 1 | 6071 | 48 | 2 | 6079 | 57 | 1 | 6083 | 6 | 4 | 6087 | 18 | 2 | 6091 | 15 | 1 |
| 6095 | 21 | 4 | 6099 | 4 | 4 | 6103 | 14 | 2 | 6107 | 15 | 2 | 6115 | 5 | 2 | 6119 | 41 | 1 |
| 6123 | 4 | 4 | 6127 | 11 | 2 | 6131 | 31 | 1 | 6135 | 18 | 4 | 6139 | 10 | 2 | 6143 | 41 | 1 |
| 6151 | 59 | 1 | 6155 | 14 | 2 | 6159 | 40 | 2 | 6163 | 11 | 1 | 6167 | 33 | 2 | 6179 | 21 | 2 |
| 6187 | 6 | 2 | 6191 | 49 | 2 | 6195 | 2 | 8 | 6199 | 39 | 1 | 6203 | 17 | 1 | 6207 | 25 | 2 |
| 6211 | 15 | 1 | 6215 | 24 | 4 | 6227 | 9 | 2 | 6231 | 16 | 4 | 6235 | 3 | 4 | 6239 | 45 | 2 |
| 6243 | 11 | 2 | 6247 | 43 | 1 | 6251 | 6 | 4 | 6259 | 5 | 2 | 6263 | 77 | 1 | 6267 | 6 | 2 |
| 6271 | 51 | 1 | 6279 | 7 | 8 | 6283 | 6 | 2 | 6287 | 51 | 1 | 6295 | 12 | 2 | 6299 | 43 | 1 |
| 6303 | 12 | 4 | 6307 | 2 | 4 | 6311 | 89 | 1 | 6315 | 4 | 4 | 6319 | 26 | 2 | 6323 | 21 | 1 |
| 6331 | 9 | 2 | 6335 | 23 | 4 | 6339 | 10 | 2 | 6343 | 33 | 1 | 6347 | 14 | 2 | 6351 | 11 | 4 |
| 6355 | 4 | 4 | 6359 | 101 | 1 | 6367 | 37 | 1 | 6371 | 18 | 2 | 6379 | 17 | 1 | 6383 | 36 | 2 |
| 6387 | 9 | 2 | 6391 | 12 | 4 | 6395 | 8 | 2 | 6403 | 5 | 2 | 6407 | 23 | 2 | 6411 | 12 | 2 |
| 6415 | 25 | 2 | 6423 | 19 | 2 | 6427 | 9 | 1 | 6431 | 57 | 2 | 6439 | 31 | 2 | 6443 | 13 | 2 |
| 6447 | 12 | 4 | 6451 | 17 | 1 | 6455 | 30 | 2 | 6459 | 15 | 2 | 6463 | 15 | 2 | 6467 | 10 | 2 |
| 6479 | 21 | 4 | 6483 | 6 | 2 | 6487 | 21 | 2 | 6491 | 31 | 1 | 6495 | 13 | 4 | 6499 | 7 | 2 |

*APPENDIX 1*

Table 1A

| Disc | H/G | G | Disc | H/G | G | Disc | H/G | G | Disc | H/G | G | Disc | H/G | G | Disc | H/G | G |
|---|---|---|---|---|---|---|---|---|---|---|---|---|---|---|---|---|---|
| 6503 | 41 | 2 | 6511 | 22 | 2 | 6515 | 13 | 2 | 6519 | 17 | 4 | 6523 | 7 | 2 | 6527 | 40 | 2 |
| 6531 | 6 | 4 | 6535 | 15 | 2 | 6539 | 14 | 2 | 6547 | 11 | 1 | 6551 | 117 | 1 | 6555 | 3 | 8 |
| 6559 | 19 | 2 | 6563 | 23 | 1 | 6567 | 8 | 4 | 6571 | 15 | 1 | 6583 | 18 | 2 | 6587 | 13 | 2 |
| 6595 | 8 | 2 | 6599 | 109 | 1 | 6603 | 3 | 4 | 6607 | 45 | 1 | 6611 | 21 | 2 | 6619 | 13 | 1 |
| 6623 | 21 | 2 | 6631 | 22 | 2 | 6635 | 15 | 2 | 6639 | 45 | 2 | 6643 | 3 | 4 | 6659 | 23 | 1 |
| 6663 | 30 | 2 | 6667 | 5 | 2 | 6671 | 44 | 2 | 6679 | 55 | 1 | 6683 | 16 | 2 | 6691 | 21 | 1 |
| 6695 | 23 | 4 | 6699 | 3 | 8 | 6703 | 23 | 1 | 6707 | 14 | 2 | 6711 | 37 | 2 | 6715 | 3 | 4 |
| 6719 | 105 | 1 | 6731 | 17 | 2 | 6735 | 11 | 4 | 6739 | 9 | 2 | 6743 | 27 | 2 | 6747 | 4 | 4 |
| 6751 | 33 | 2 | 6755 | 5 | 4 | 6763 | 9 | 1 | 6767 | 43 | 2 | 6771 | 4 | 4 | 6779 | 39 | 1 |
| 6783 | 7 | 8 | 6787 | 6 | 2 | 6791 | 81 | 1 | 6799 | 22 | 2 | 6803 | 19 | 1 | 6807 | 20 | 2 |
| 6815 | 23 | 4 | 6819 | 11 | 2 | 6823 | 33 | 1 | 6827 | 17 | 1 | 6835 | 9 | 2 | 6839 | 54 | 2 |
| 6843 | 6 | 2 | 6847 | 13 | 4 | 6851 | 9 | 4 | 6855 | 15 | 4 | 6863 | 81 | 1 | 6871 | 45 | 1 |
| 6879 | 32 | 2 | 6883 | 9 | 1 | 6887 | 39 | 2 | 6891 | 13 | 2 | 6895 | 7 | 2 | 6899 | 35 | 1 |
| 6907 | 17 | 1 | 6911 | 87 | 1 | 6915 | 7 | 4 | 6919 | 17 | 4 | 6923 | 4 | 4 | 6927 | 19 | 2 |
| 6931 | 6 | 2 | 6935 | 22 | 4 | 6943 | 24 | 2 | 6947 | 29 | 1 | 6951 | 20 | 4 | 6955 | 3 | 4 |
| 6959 | 95 | 1 | 6963 | 3 | 4 | 6967 | 33 | 2 | 6971 | 45 | 1 | 6979 | 7 | 2 | 6983 | 57 | 1 |
| 6987 | 3 | 4 | 6991 | 71 | 1 | 6995 | 16 | 2 | 6999 | 33 | 2 | 7003 | 8 | 2 | 7015 | 7 | 4 |
| 7019 | 43 | 1 | 7023 | 22 | 2 | 7027 | 11 | 1 | 7031 | 54 | 2 | 7035 | 2 | 8 | 7039 | 43 | 1 |
| 7043 | 23 | 1 | 7051 | 8 | 2 | 7055 | 23 | 4 | 7059 | 8 | 4 | 7063 | 20 | 2 | 7067 | 7 | 2 |
| 7071 | 35 | 2 | 7079 | 85 | 1 | 7087 | 15 | 2 | 7091 | 19 | 2 | 7095 | 7 | 8 | 7099 | 7 | 2 |
| 7103 | 77 | 1 | 7107 | 3 | 2 | 7111 | 26 | 2 | 7115 | 15 | 2 | 7123 | 5 | 2 | 7127 | 79 | 1 |
| 7131 | 10 | 2 | 7135 | 19 | 2 | 7143 | 23 | 2 | 7147 | 7 | 2 | 7151 | 85 | 1 | 7159 | 65 | 1 |
| 7163 | 5 | 4 | 7167 | 32 | 2 | 7171 | 10 | 2 | 7179 | 11 | 2 | 7183 | 16 | 2 | 7187 | 25 | 1 |
| 7195 | 8 | 2 | 7199 | 57 | 1 | 7207 | 29 | 1 | 7211 | 35 | 1 | 7215 | 9 | 8 | 7219 | 15 | 1 |
| 7223 | 42 | 2 | 7231 | 22 | 2 | 7235 | 11 | 2 | 7239 | 12 | 4 | 7243 | 13 | 1 | 7247 | 47 | 1 |
| 7251 | 17 | 2 | 7255 | 30 | 2 | 7259 | 9 | 4 | 7271 | 58 | 2 | 7279 | 35 | 2 | 7283 | 25 | 1 |
| 7287 | 9 | 4 | 7291 | 6 | 2 | 7295 | 40 | 2 | 7303 | 15 | 2 | 7307 | 25 | 1 | 7311 | 44 | 2 |
| 7315 | 2 | 8 | 7319 | 44 | 2 | 7323 | 9 | 2 | 7327 | 17 | 2 | 7331 | 33 | 1 | 7339 | 9 | 2 |
| 7343 | 37 | 2 | 7347 | 4 | 4 | 7351 | 33 | 4 | 7355 | 18 | 2 | 7359 | 18 | 4 | 7363 | 5 | 2 |
| 7367 | 47 | 2 | 7379 | 14 | 4 | 7383 | 14 | 4 | 7387 | 5 | 2 | 7391 | 60 | 2 | 7395 | 2 | 8 |
| 7403 | 11 | 2 | 7411 | 25 | 1 | 7415 | 29 | 2 | 7419 | 12 | 2 | 7423 | 20 | 2 | 7427 | 14 | 2 |
| 7431 | 35 | 2 | 7435 | 5 | 2 | 7439 | 58 | 2 | 7447 | 21 | 2 | 7451 | 35 | 1 | 7455 | 7 | 8 |
| 7459 | 15 | 1 | 7463 | 33 | 2 | 7467 | 5 | 4 | 7471 | 29 | 2 | 7483 | 5 | 2 | 7487 | 65 | 1 |
| 7491 | 4 | 2 | 7495 | 24 | 2 | 7499 | 33 | 1 | 7503 | 10 | 4 | 7507 | 11 | 1 | 7511 | 24 | 4 |
| 7519 | 25 | 2 | 7523 | 35 | 1 | 7527 | 10 | 4 | 7531 | 12 | 2 | 7535 | 24 | 4 | 7539 | 7 | 4 |
| 7543 | 18 | 2 | 7547 | 15 | 1 | 7555 | 6 | 2 | 7559 | 115 | 1 | 7563 | 12 | 2 | 7567 | 11 | 4 |
| 7571 | 9 | 2 | 7579 | 4 | 4 | 7583 | 63 | 1 | 7591 | 65 | 1 | 7599 | 13 | 4 | 7603 | 11 | 1 |
| 7607 | 89 | 1 | 7611 | 5 | 4 | 7615 | 25 | 2 | 7619 | 23 | 2 | 7627 | 5 | 2 | 7631 | 53 | 2 |
| 7635 | 5 | 4 | 7639 | 31 | 4 | 7643 | 29 | 2 | 7647 | 33 | 2 | 7651 | 10 | 2 | 7655 | 38 | 2 |
| 7663 | 22 | 2 | 7667 | 5 | 4 | 7671 | 42 | 2 | 7679 | 53 | 2 | 7683 | 3 | 4 | 7687 | 29 | 1 |
| 7691 | 43 | 1 | 7699 | 27 | 1 | 7703 | 81 | 1 | 7707 | 4 | 4 | 7711 | 17 | 2 | 7715 | 9 | 2 |
| 7719 | 24 | 4 | 7723 | 9 | 1 | 7727 | 81 | 1 | 7735 | 6 | 8 | 7739 | 20 | 2 | 7743 | 13 | 4 |
| 7747 | 8 | 2 | 7751 | 55 | 2 | 7755 | 2 | 8 | 7759 | 49 | 1 | 7763 | 11 | 2 | 7771 | 9 | 2 |
| 7779 | 12 | 2 | 7783 | 22 | 2 | 7787 | 14 | 2 | 7795 | 8 | 2 | 7799 | 48 | 2 | 7807 | 18 | 2 |
| 7811 | 21 | 2 | 7815 | 13 | 4 | 7819 | 8 | 2 | 7823 | 75 | 1 | 7827 | 9 | 2 | 7831 | 33 | 2 |
| 7835 | 15 | 2 | 7843 | 4 | 4 | 7847 | 14 | 4 | 7851 | 10 | 2 | 7855 | 22 | 2 | 7859 | 17 | 2 |
| 7863 | 17 | 2 | 7867 | 11 | 1 | 7871 | 60 | 2 | 7879 | 49 | 1 | 7883 | 17 | 1 | 7887 | 18 | 4 |
| 7891 | 6 | 2 | 7895 | 56 | 2 | 7899 | 11 | 2 | 7903 | 22 | 2 | 7907 | 21 | 1 | 7915 | 7 | 2 |
| 7919 | 97 | 1 | 7923 | 4 | 4 | 7927 | 47 | 1 | 7931 | 11 | 4 | 7939 | 10 | 2 | 7951 | 51 | 1 |
| 7955 | 11 | 4 | 7959 | 20 | 4 | 7963 | 13 | 1 | 7967 | 22 | 2 | 7971 | 15 | 2 | 7979 | 22 | 2 |
| 7991 | 50 | 2 | 7995 | 2 | 8 | 7999 | 37 | 2 | 8003 | 13 | 2 | 8007 | 16 | 4 | 8011 | 25 | 1 |
| 8015 | 23 | 4 | 8023 | 11 | 2 | 8027 | 14 | 2 | 8031 | 30 | 2 | 8035 | 7 | 2 | 8039 | 113 | 1 |
| 8043 | 4 | 4 | 8047 | 17 | 2 | 8051 | 17 | 2 | 8059 | 21 | 1 | 8063 | 43 | 2 | 8067 | 10 | 2 |
| 8071 | 33 | 2 | 8079 | 25 | 2 | 8083 | 8 | 2 | 8087 | 81 | 1 | 8095 | 28 | 2 | 8099 | 7 | 4 |
| 8103 | 12 | 4 | 8111 | 121 | 1 | 8115 | 6 | 4 | 8119 | 26 | 2 | 8123 | 21 | 1 | 8131 | 6 | 2 |

Table 1A

| Disc | H/G | G | Disc | H/G | G | Disc | H/G | G | Disc | H/G | G | Disc | H/G | G | Disc | H/G | G |
|---|---|---|---|---|---|---|---|---|---|---|---|---|---|---|---|---|---|
| 8135 | 47 | 2 | 8139 | 18 | 2 | 8143 | 11 | 2 | 8147 | 37 | 1 | 8151 | 11 | 8 | 8155 | 3 | 4 |
| 8159 | 53 | 2 | 8167 | 33 | 1 | 8171 | 21 | 1 | 8179 | 25 | 1 | 8187 | 7 | 2 | 8191 | 55 | 1 |
| 8195 | 8 | 4 | 8203 | 9 | 2 | 8207 | 48 | 2 | 8211 | 3 | 8 | 8215 | 9 | 4 | 8219 | 35 | 1 |
| 8223 | 29 | 2 | 8227 | 5 | 2 | 8231 | 107 | 1 | 8239 | 16 | 4 | 8243 | 21 | 1 | 8247 | 20 | 2 |
| 8251 | 10 | 2 | 8255 | 17 | 4 | 8259 | 15 | 2 | 8263 | 43 | 1 | 8267 | 15 | 2 | 8279 | 63 | 2 |
| 8283 | 4 | 4 | 8287 | 45 | 1 | 8291 | 47 | 1 | 8295 | 8 | 8 | 8299 | 8 | 2 | 8311 | 61 | 1 |
| 8315 | 19 | 2 | 8319 | 20 | 4 | 8323 | 3 | 4 | 8327 | 26 | 2 | 8331 | 9 | 2 | 8335 | 25 | 2 |
| 8339 | 13 | 2 | 8347 | 6 | 2 | 8351 | 59 | 2 | 8355 | 5 | 4 | 8359 | 29 | 2 | 8363 | 35 | 1 |
| 8367 | 15 | 2 | 8371 | 11 | 2 | 8383 | 13 | 2 | 8387 | 21 | 1 | 8391 | 38 | 2 | 8395 | 3 | 4 |
| 8399 | 67 | 2 | 8403 | 9 | 2 | 8407 | 18 | 2 | 8411 | 20 | 2 | 8419 | 19 | 1 | 8423 | 83 | 1 |
| 8431 | 59 | 1 | 8435 | 7 | 4 | 8439 | 17 | 4 | 8443 | 11 | 1 | 8447 | 99 | 1 | 8455 | 14 | 4 |
| 8459 | 21 | 2 | 8463 | 6 | 8 | 8467 | 15 | 1 | 8471 | 40 | 2 | 8479 | 25 | 2 | 8483 | 15 | 2 |
| 8491 | 10 | 2 | 8495 | 32 | 2 | 8499 | 12 | 2 | 8503 | 24 | 2 | 8507 | 9 | 2 | 8511 | 47 | 2 |
| 8515 | 4 | 4 | 8519 | 55 | 2 | 8527 | 43 | 1 | 8531 | 25 | 2 | 8535 | 13 | 4 | 8539 | 17 | 1 |
| 8543 | 97 | 1 | 8547 | 2 | 8 | 8551 | 25 | 2 | 8555 | 8 | 4 | 8563 | 9 | 1 | 8567 | 42 | 2 |
| 8571 | 18 | 2 | 8579 | 21 | 2 | 8583 | 25 | 2 | 8587 | 9 | 2 | 8599 | 63 | 1 | 8603 | 12 | 2 |
| 8607 | 10 | 4 | 8611 | 7 | 2 | 8615 | 59 | 2 | 8623 | 51 | 1 | 8627 | 21 | 1 | 8635 | 4 | 4 |
| 8639 | 44 | 2 | 8643 | 4 | 4 | 8647 | 31 | 1 | 8651 | 19 | 2 | 8655 | 21 | 4 | 8659 | 11 | 2 |
| 8663 | 67 | 1 | 8671 | 16 | 4 | 8679 | 22 | 4 | 8683 | 8 | 2 | 8687 | 21 | 4 | 8691 | 13 | 2 |
| 8695 | 9 | 4 | 8699 | 35 | 1 | 8707 | 15 | 1 | 8711 | 66 | 2 | 8715 | 2 | 8 | 8719 | 53 | 1 |
| 8723 | 7 | 4 | 8727 | 25 | 2 | 8731 | 17 | 1 | 8735 | 53 | 2 | 8743 | 13 | 2 | 8747 | 21 | 1 |
| 8751 | 36 | 2 | 8755 | 5 | 4 | 8759 | 54 | 2 | 8763 | 6 | 4 | 8767 | 18 | 2 | 8779 | 15 | 1 |
| 8783 | 73 | 1 | 8787 | 3 | 4 | 8791 | 41 | 2 | 8795 | 16 | 2 | 8799 | 20 | 4 | 8803 | 9 | 1 |
| 8807 | 81 | 1 | 8815 | 12 | 4 | 8819 | 49 | 1 | 8823 | 15 | 4 | 8827 | 3 | 4 | 8831 | 109 | 1 |
| 8835 | 2 | 8 | 8839 | 77 | 1 | 8843 | 13 | 2 | 8851 | 11 | 2 | 8855 | 11 | 8 | 8859 | 8 | 2 |
| 8863 | 29 | 1 | 8867 | 27 | 2 | 8871 | 23 | 2 | 8879 | 65 | 2 | 8887 | 43 | 1 | 8891 | 16 | 2 |
| 8895 | 23 | 4 | 8899 | 7 | 2 | 8903 | 35 | 2 | 8907 | 11 | 2 | 8911 | 11 | 4 | 8915 | 11 | 2 |
| 8923 | 19 | 1 | 8927 | 41 | 2 | 8931 | 8 | 4 | 8935 | 23 | 2 | 8939 | 19 | 2 | 8943 | 15 | 4 |
| 8947 | 5 | 2 | 8951 | 135 | 1 | 8963 | 29 | 1 | 8971 | 19 | 1 | 8979 | 6 | 4 | 8983 | 19 | 2 |
| 8987 | 8 | 4 | 8995 | 5 | 4 | 8999 | 99 | 1 | 9003 | 6 | 2 | 9007 | 35 | 1 | 9011 | 33 | 1 |
| 9015 | 18 | 4 | 9019 | 15 | 2 | 9023 | 40 | 2 | 9031 | 27 | 2 | 9035 | 10 | 4 | 9039 | 20 | 4 |
| 9043 | 15 | 1 | 9047 | 44 | 2 | 9051 | 5 | 4 | 9055 | 18 | 2 | 9059 | 39 | 1 | 9067 | 9 | 1 |
| 9071 | 69 | 2 | 9079 | 26 | 2 | 9083 | 12 | 2 | 9087 | 8 | 4 | 9091 | 21 | 1 | 9095 | 19 | 4 |
| 9103 | 57 | 1 | 9107 | 13 | 2 | 9111 | 32 | 2 | 9115 | 7 | 2 | 9119 | 62 | 2 | 9123 | 9 | 2 |
| 9127 | 57 | 1 | 9131 | 20 | 2 | 9139 | 3 | 4 | 9143 | 36 | 2 | 9147 | 10 | 2 | 9151 | 67 | 1 |
| 9155 | 18 | 2 | 9159 | 24 | 4 | 9167 | 27 | 2 | 9179 | 19 | 2 | 9183 | 20 | 2 | 9187 | 21 | 1 |
| 9191 | 31 | 4 | 9195 | 5 | 4 | 9199 | 51 | 1 | 9203 | 31 | 1 | 9211 | 9 | 2 | 9215 | 29 | 4 |
| 9219 | 4 | 2 | 9223 | 17 | 2 | 9227 | 25 | 1 | 9231 | 25 | 4 | 9235 | 7 | 2 | 9239 | 139 | 1 |
| 9247 | 27 | 2 | 9255 | 13 | 2 | 9259 | 12 | 2 | 9263 | 31 | 2 | 9267 | 11 | 2 | 9271 | 30 | 2 |
| 9283 | 11 | 1 | 9287 | 39 | 2 | 9291 | 8 | 4 | 9299 | 25 | 2 | 9303 | 17 | 4 | 9307 | 5 | 2 |
| 9311 | 97 | 1 | 9319 | 41 | 1 | 9323 | 29 | 1 | 9327 | 28 | 2 | 9331 | 5 | 4 | 9335 | 49 | 2 |
| 9339 | 5 | 4 | 9343 | 51 | 1 | 9347 | 12 | 2 | 9355 | 6 | 2 | 9363 | 10 | 2 | 9367 | 7 | 4 |
| 9371 | 49 | 1 | 9379 | 12 | 2 | 9383 | 51 | 2 | 9391 | 55 | 1 | 9395 | 12 | 2 | 9399 | 24 | 4 |
| 9403 | 11 | 1 | 9407 | 46 | 2 | 9411 | 15 | 2 | 9415 | 13 | 4 | 9419 | 35 | 1 | 9427 | 7 | 2 |
| 9431 | 91 | 1 | 9435 | 4 | 8 | 9439 | 75 | 1 | 9443 | 5 | 4 | 9447 | 11 | 4 | 9451 | 12 | 2 |
| 9455 | 30 | 4 | 9463 | 45 | 1 | 9467 | 41 | 1 | 9471 | 9 | 8 | 9479 | 101 | 1 | 9483 | 4 | 4 |
| 9487 | 19 | 2 | 9491 | 45 | 1 | 9499 | 7 | 4 | 9503 | 20 | 4 | 9507 | 8 | 2 | 9511 | 69 | 1 |
| 9515 | 7 | 4 | 9519 | 24 | 1 | 9523 | 6 | 2 | 9527 | 33 | 2 | 9535 | 17 | 2 | 9539 | 55 | 1 |
| 9543 | 26 | 2 | 9547 | 13 | 1 | 9551 | 129 | 1 | 9563 | 9 | 2 | 9571 | 10 | 2 | 9579 | 8 | 4 |
| 9587 | 23 | 1 | 9591 | 20 | 4 | 9595 | 4 | 4 | 9599 | 49 | 2 | 9607 | 21 | 2 | 9611 | 22 | 2 |
| 9615 | 9 | 4 | 9619 | 19 | 1 | 9623 | 95 | 1 | 9627 | 9 | 2 | 9631 | 77 | 1 | 9635 | 11 | 4 |
| 9643 | 11 | 4 | 9647 | 43 | 2 | 9651 | 14 | 2 | 9655 | 26 | 2 | 9659 | 19 | 2 | 9663 | 27 | 2 |
| 9667 | 6 | 2 | 9671 | 63 | 1 | 9679 | 71 | 2 | 9683 | 9 | 2 | 9687 | 30 | 2 | 9691 | 10 | 2 |
| 9695 | 23 | 4 | 9699 | 7 | 4 | 9703 | 17 | 2 | 9707 | 13 | 2 | 9715 | 6 | 4 | 9719 | 133 | 1 |
| 9723 | 6 | 4 | 9727 | 19 | 2 | 9731 | 16 | 2 | 9735 | 8 | 8 | 9739 | 13 | 1 | 9743 | 105 | 1 |

Table 1A

| Disc | H/G | G | Disc | H/G | G | Disc | H/G | G | Disc | H/G | G | Disc | H/G | G | Disc | H/G | G |
|------|-----|---|------|-----|---|------|-----|---|------|-----|---|------|-----|---|------|-----|---|
| 9755 | 10 | 2 | 9759 | 46 | 2 | 9763 | 8 | 2 | 9767 | 89 | 1 | 9771 | 13 | 2 | 9779 | 12 | 4 |
| 9787 | 11 | 1 | 9791 | 119 | 1 | 9795 | 5 | 4 | 9799 | 33 | 2 | 9803 | 37 | 1 | 9807 | 16 | 4 |
| 9811 | 21 | 1 | 9815 | 19 | 4 | 9823 | 6 | 4 | 9827 | 14 | 2 | 9831 | 27 | 4 | 9835 | 4 | 4 |
| 9839 | 91 | 1 | 9843 | 3 | 4 | 9847 | 22 | 2 | 9851 | 45 | 1 | 9859 | 21 | 1 | 9863 | 48 | 2 |
| 9867 | 2 | 8 | 9871 | 49 | 1 | 9879 | 19 | 4 | 9883 | 17 | 1 | 9887 | 75 | 1 | 9895 | 28 | 2 |
| 9899 | 15 | 2 | 9903 | 28 | 2 | 9907 | 15 | 1 | 9911 | 34 | 4 | 9915 | 6 | 4 | 9919 | 17 | 4 |
| 9923 | 25 | 1 | 9931 | 23 | 2 | 9935 | 47 | 2 | 9939 | 14 | 2 | 9943 | 28 | 2 | 9951 | 19 | 4 |
| 9955 | 4 | 4 | 9959 | 65 | 2 | 9967 | 39 | 1 | 9979 | 10 | 2 | 9983 | 46 | 2 | 9987 | 13 | 2 |
| 9991 | 16 | 2 | 9995 | 20 | 2 | | | | | | | | | | | | |

Table 1B

| Disc | H/G | G | Disc | H/G | G | Disc | H/G | G | Disc | H/G | G | Disc | H/G | G | Disc | H/G | G |
|---|---|---|---|---|---|---|---|---|---|---|---|---|---|---|---|---|---|
| 4 | 1 | 1 | 8 | 1 | 1 | 20 | 1 | 2 | 24 | 1 | 2 | 40 | 1 | 2 | 52 | 1 | 2 |
| 56 | 2 | 2 | 68 | 2 | 2 | 84 | 1 | 4 | 88 | 1 | 2 | 104 | 3 | 2 | 116 | 3 | 2 |
| 120 | 1 | 4 | 132 | 1 | 4 | 136 | 2 | 2 | 148 | 1 | 2 | 152 | 3 | 2 | 164 | 4 | 2 |
| 168 | 1 | 4 | 184 | 2 | 2 | 212 | 3 | 2 | 228 | 1 | 4 | 232 | 1 | 2 | 244 | 3 | 2 |
| 248 | 4 | 2 | 260 | 2 | 4 | 264 | 2 | 4 | 276 | 2 | 4 | 280 | 1 | 4 | 292 | 2 | 2 |
| 296 | 5 | 2 | 308 | 2 | 4 | 312 | 1 | 4 | 328 | 2 | 2 | 340 | 1 | 4 | 344 | 5 | 2 |
| 356 | 6 | 2 | 372 | 2 | 2 | 376 | 4 | 2 | 388 | 2 | 2 | 404 | 7 | 2 | 408 | 1 | 4 |
| 420 | 1 | 8 | 424 | 3 | 2 | 436 | 3 | 2 | 440 | 3 | 4 | 452 | 4 | 2 | 456 | 2 | 4 |
| 472 | 3 | 2 | 488 | 5 | 2 | 516 | 3 | 4 | 520 | 1 | 4 | 532 | 1 | 4 | 536 | 7 | 2 |
| 548 | 4 | 2 | 552 | 2 | 4 | 564 | 2 | 4 | 568 | 2 | 2 | 580 | 2 | 4 | 584 | 8 | 2 |
| 596 | 7 | 2 | 616 | 2 | 4 | 628 | 3 | 2 | 632 | 4 | 2 | 644 | 4 | 4 | 660 | 1 | 8 |
| 664 | 5 | 2 | 680 | 3 | 4 | 692 | 7 | 2 | 696 | 3 | 4 | 708 | 1 | 4 | 712 | 4 | 2 |
| 724 | 5 | 2 | 728 | 3 | 4 | 740 | 4 | 4 | 744 | 3 | 4 | 760 | 1 | 4 | 772 | 2 | 2 |
| 776 | 10 | 2 | 788 | 5 | 2 | 804 | 3 | 4 | 808 | 3 | 2 | 820 | 2 | 4 | 824 | 10 | 2 |
| 836 | 5 | 4 | 840 | 1 | 8 | 852 | 2 | 4 | 856 | 3 | 2 | 868 | 2 | 4 | 872 | 5 | 2 |
| 884 | 4 | 4 | 888 | 3 | 4 | 904 | 4 | 2 | 916 | 5 | 2 | 920 | 5 | 2 | 932 | 6 | 2 |
| 948 | 3 | 4 | 952 | 2 | 4 | 964 | 6 | 2 | 984 | 3 | 4 | 996 | 3 | 4 | 1012 | 1 | 4 |
| 1016 | 8 | 2 | 1028 | 8 | 2 | 1032 | 2 | 4 | 1048 | 3 | 2 | 1060 | 2 | 4 | 1064 | 5 | 4 |
| 1076 | 11 | 2 | 1092 | 1 | 8 | 1096 | 6 | 2 | 1108 | 3 | 2 | 1112 | 7 | 2 | 1124 | 10 | 2 |
| 1128 | 2 | 4 | 1140 | 2 | 8 | 1144 | 3 | 4 | 1160 | 5 | 4 | 1172 | 9 | 2 | 1192 | 3 | 2 |
| 1204 | 2 | 4 | 1208 | 6 | 2 | 1220 | 4 | 4 | 1236 | 3 | 4 | 1240 | 2 | 4 | 1252 | 4 | 2 |
| 1256 | 13 | 2 | 1268 | 5 | 2 | 1272 | 3 | 4 | 1284 | 5 | 4 | 1288 | 2 | 4 | 1304 | 11 | 2 |
| 1316 | 6 | 4 | 1320 | 1 | 8 | 1336 | 6 | 2 | 1348 | 4 | 2 | 1364 | 7 | 4 | 1380 | 1 | 8 |
| 1384 | 5 | 2 | 1396 | 7 | 2 | 1412 | 8 | 2 | 1416 | 4 | 4 | 1428 | 1 | 8 | 1432 | 3 | 2 |
| 1448 | 9 | 2 | 1460 | 5 | 4 | 1464 | 3 | 4 | 1480 | 3 | 4 | 1492 | 5 | 2 | 1496 | 7 | 4 |
| 1508 | 4 | 4 | 1524 | 5 | 4 | 1528 | 4 | 2 | 1540 | 1 | 8 | 1544 | 10 | 2 | 1556 | 11 | 2 |
| 1560 | 2 | 8 | 1572 | 3 | 4 | 1576 | 5 | 2 | 1588 | 3 | 2 | 1592 | 10 | 2 | 1604 | 10 | 2 |
| 1608 | 4 | 4 | 1624 | 4 | 4 | 1636 | 8 | 2 | 1640 | 4 | 4 | 1652 | 5 | 4 | 1668 | 3 | 4 |
| 1672 | 2 | 4 | 1684 | 5 | 2 | 1688 | 5 | 2 | 1704 | 6 | 4 | 1716 | 2 | 8 | 1720 | 3 | 4 |
| 1732 | 6 | 2 | 1736 | 6 | 4 | 1748 | 5 | 4 | 1752 | 2 | 4 | 1768 | 2 | 4 | 1780 | 2 | 4 |
| 1784 | 16 | 2 | 1796 | 10. | 2 | 1812 | 3 | 4 | 1816 | 7 | 2 | 1828 | 4 | 2 | 1832 | 13 | 2 |
| 1844 | 15 | 2 | 1848 | 1 | 8 | 1860 | 2 | 8 | 1864 | 4 | 2 | 1876 | 4 | 4 | 1880 | 5 | 4 |
| 1892 | 3 | 4 | 1896 | 5 | 4 | 1912 | 4 | 2 | 1924 | 4 | 4 | 1928 | 10 | 2 | 1940 | 5 | 4 |
| 1956 | 5 | 4 | 1972 | 3 | 4 | 1976 | 7 | 4 | 1988 | 6 | 4 | 1992 | 2 | 4 | 2004 | 4 | 4 |
| 2008 | 7 | 2 | 2020 | 2 | 4 | 2024 | 7 | 4 | 2036 | 15 | 2 | 2040 | 2 | 8 | 2056 | 8 | 2 |
| 2068 | 3 | 4 | 2072 | 4 | 4 | 2084 | 16 | 2 | 2104 | 6 | 2 | 2120 | 7 | 4 | 2132 | 3 | 4 |
| 2136 | 5 | 4 | 2148 | 3 | 4 | 2152 | 5 | 2 | 2164 | 5 | 4 | 2168 | 12 | 2 | 2180 | 8 | 4 |
| 2184 | 3 | 8 | 2212 | 2 | 4 | 2216 | 11 | 2 | 2228 | 9 | 2 | 2244 | 2 | 8 | 2248 | 4 | 2 |
| 2260 | 3 | 4 | 2264 | 15 | 2 | 2276 | 16 | 2 | 2280 | 2 | 8 | 2292 | 4 | 4 | 2296 | 4 | 4 |
| 2308 | 4 | 2 | 2324 | 7 | 4 | 2328 | 4 | 4 | 2344 | 9 | 2 | 2356 | 4 | 4 | 2360 | 5 | 4 |
| 2372 | 12 | 2 | 2388 | 3 | 4 | 2392 | 2 | 4 | 2404 | 10 | 2 | 2408 | 6 | 4 | 2424 | 3 | 4 |
| 2436 | 2 | 8 | 2440 | 3 | 4 | 2452 | 5 | 2 | 2456 | 17 | 2 | 2468 | 6 | 4 | 2472 | 3 | 4 |
| 2488 | 6 | 2 | 2504 | 18 | 2 | 2516 | 9 | 4 | 2532 | 5 | 4 | 2536 | 7 | 2 | 2552 | 5 | 4 |
| 2564 | 14 | 2 | 2568 | 4 | 4 | 2580 | 2 | 8 | 2584 | 4 | 4 | 2596 | 5 | 4 | 2612 | 7 | 2 |
| 2616 | 7 | 4 | 2632 | 2 | 4 | 2644 | 9 | 2 | 2648 | 11 | 2 | 2660 | 3 | 8 | 2676 | 3 | 4 |
| 2680 | 3 | 4 | 2692 | 6 | 2 | 2696 | 12 | 2 | 2708 | 15 | 2 | 2712 | 5 | 4 | 2724 | 5 | 4 |
| 2728 | 3 | 4 | 2740 | 3 | 4 | 2756 | 10 | 2 | 2760 | 2 | 8 | 2776 | 5 | 2 | 2788 | 2 | 4 |
| 2792 | 13 | 2 | 2804 | 17 | 2 | 2820 | 3 | 8 | 2824 | 12 | 2 | 2836 | 5 | 2 | 2840 | 8 | 4 |
| 2852 | 6 | 4 | 2856 | 3 | 8 | 2868 | 4 | 4 | 2872 | 6 | 2 | 2884 | 4 | 4 | 2920 | 3 | 4 |
| 2932 | 7 | 2 | 2936 | 20 | 2 | 2948 | 5 | 4 | 2964 | 3 | 8 | 2968 | 2 | 4 | 2980 | 4 | 4 |
| 2984 | 13 | 2 | 2996 | 8 | 4 | 3012 | 3 | 4 | 3016 | 5 | 4 | 3028 | 5 | 2 | 3032 | 11 | 2 |
| 3044 | 20 | 2 | 3048 | 3 | 4 | 3064 | 12 | 2 | 3076 | 10 | 2 | 3080 | 4 | 8 | 3092 | 13 | 2 |
| 3108 | 2 | 8 | 3112 | 7 | 2 | 3124 | 5 | 4 | 3128 | 6 | 4 | 3140 | 4 | 4 | 3144 | 4 | 4 |
| 3156 | 8 | 4 | 3160 | 4 | 4 | 3172 | 2 | 4 | 3176 | 21 | 2 | 3188 | 15 | 2 | 3192 | 2 | 8 |
| 3208 | 6 | 2 | 3220 | 2 | 8 | 3224 | 7 | 4 | 3236 | 16 | 2 | 3252 | 3 | 4 | 3256 | 3 | 4 |

Table 1B

| Disc | H/G | G | Disc | H/G | G | Disc | H/G | G | Disc | H/G | G | Disc | H/G | G | Disc | H/G | G |
|---|---|---|---|---|---|---|---|---|---|---|---|---|---|---|---|---|---|
| 3268 | 3 | 4 | 3272 | 14 | 2 | 3284 | 15 | 2 | 3288 | 5 | 4 | 3304 | 3 | 4 | 3316 | 11 | 2 |
| 3320 | 5 | 4 | 3336 | 4 | 4 | 3352 | 7 | 2 | 3368 | 13 | 2 | 3396 | 7 | 4 | 3412 | 5 | 2 |
| 3416 | 11 | 4 | 3428 | 16 | 2 | 3432 | 2 | 8 | 3444 | 3 | 8 | 3448 | 4 | 2 | 3460 | 4 | 4 |
| 3464 | 22 | 2 | 3476 | 8 | 4 | 3480 | 2 | 8 | 3496 | 5 | 4 | 3508 | 5 | 2 | 3512 | 10 | 2 |
| 3524 | 20 | 2 | 3540 | 3 | 8 | 3544 | 9 | 2 | 3556 | 4 | 4 | 3560 | 6 | 4 | 3572 | 7 | 4 |
| 3576 | 7 | 4 | 3588 | 2 | 8 | 3592 | 6 | 2 | 3604 | 6 | 4 | 3608 | 7 | 4 | 3620 | 6 | 4 |
| 3624 | 7 | 4 | 3640 | 2 | 8 | 3652 | 3 | 4 | 3656 | 18 | 2 | 3668 | 5 | 4 | 3684 | 5 | 4 |
| 3688 | 9 | 2 | 3704 | 20 | 2 | 3716 | 18 | 2 | 3720 | 3 | 8 | 3732 | 4 | 4 | 3736 | 13 | 2 |
| 3748 | 10 | 2 | 3752 | 4 | 4 | 3764 | 23 | 2 | 3768 | 3 | 4 | 3784 | 4 | 4 | 3796 | 3 | 4 |
| 3812 | 16 | 2 | 3828 | 2 | 8 | 3832 | 8 | 2 | 3848 | 7 | 4 | 3860 | 11 | 2 | 3864 | 3 | 8 |
| 3876 | 3 | 8 | 3880 | 3 | 4 | 3892 | 3 | 4 | 3896 | 18 | 2 | 3908 | 10 | 2 | 3912 | 6 | 4 |
| 3928 | 5 | 2 | 3940 | 6 | 4 | 3944 | 11 | 4 | 3956 | 9 | 4 | 3972 | 3 | 4 | 3976 | 4 | 4 |
| 3988 | 7 | 2 | 3992 | 13 | 2 | 4004 | 5 | 8 | 4008 | 4 | 4 | 4020 | 2 | 8 | 4024 | 10 | 2 |
| 4036 | 10 | 2 | 4040 | 7 | 4 | 4052 | 13 | 2 | 4072 | 9 | 2 | 4084 | 11 | 2 | 4088 | 8 | 4 |
| 4120 | 3 | 4 | 4132 | 6 | 2 | 4136 | 11 | 4 | 4148 | 5 | 4 | 4152 | 3 | 4 | 4164 | 9 | 4 |
| 4168 | 6 | 2 | 4180 | 2 | 8 | 4184 | 21 | 2 | 4196 | 22 | 2 | 4216 | 4 | 4 | 4228 | 4 | 4 |
| 4244 | 13 | 2 | 4260 | 2 | 8 | 4264 | 5 | 4 | 4276 | 15 | 2 | 4280 | 9 | 4 | 4292 | 6 | 4 |
| 4296 | 8 | 4 | 4308 | 6 | 4 | 4324 | 4 | 4 | 4328 | 11 | 2 | 4340 | 4 | 8 | 4344 | 7 | 4 |
| 4360 | 3 | 4 | 4372 | 5 | 2 | 4376 | 13 | 2 | 4388 | 18 | 2 | 4404 | 7 | 4 | 4408 | 5 | 4 |
| 4420 | 2 | 8 | 4424 | 12 | 4 | 4436 | 25 | 2 | 4440 | 2 | 8 | 4452 | 2 | 8 | 4456 | 11 | 2 |
| 4468 | 7 | 2 | 4472 | 9 | 4 | 4484 | 11 | 4 | 4488 | 2 | 8 | 4504 | 11 | 2 | 4516 | 8 | 2 |
| 4520 | 11 | 4 | 4532 | 7 | 4 | 4548 | 5 | 4 | 4552 | 6 | 2 | 4564 | 6 | 4 | 4568 | 9 | 2 |
| 4580 | 6 | 4 | 4584 | 8 | 4 | 4596 | 4 | 4 | 4612 | 8 | 2 | 4616 | 28 | 2 | 4628 | 7 | 4 |
| 4632 | 6 | 4 | 4648 | 3 | 4 | 4660 | 5 | 4 | 4664 | 9 | 4 | 4676 | 12 | 4 | 4692 | 3 | 8 |
| 4696 | 15 | 2 | 4708 | 3 | 4 | 4712 | 4 | 4 | 4724 | 23 | 2 | 4728 | 5 | 4 | 4740 | 2 | 8 |
| 4744 | 12 | 2 | 4756 | 5 | 4 | 4760 | 5 | 8 | 4772 | 18 | 2 | 4776 | 7 | 4 | 4792 | 6 | 2 |
| 4804 | 8 | 2 | 4808 | 12 | 2 | 4820 | 10 | 4 | 4836 | 5 | 8 | 4852 | 5 | 2 | 4856 | 20 | 2 |
| 4868 | 16 | 2 | 4872 | 3 | 8 | 4884 | 4 | 8 | 4888 | 3 | 4 | 4904 | 21 | 2 | 4916 | 19 | 2 |
| 4920 | 3 | 8 | 4936 | 12 | 2 | 4948 | 7 | 2 | 4952 | 21 | 2 | 4964 | 8 | 4 | 4980 | 4 | 8 |
| 4984 | 4 | 4 | 4996 | 16 | 2 | 5012 | 8 | 4 | 5016 | 3 | 8 | 5028 | 5 | 4 | 5032 | 3 | 4 |
| 5044 | 5 | 4 | 5048 | 18 | 2 | 5060 | 5 | 8 | 5064 | 8 | 4 | 5080 | 5 | 4 | 5092 | 5 | 4 |
| 5108 | 17 | 2 | 5124 | 3 | 8 | 5128 | 6 | 2 | 5140 | 3 | 4 | 5144 | 29 | 2 | 5156 | 18 | 2 |
| 5160 | 2 | 8 | 5172 | 6 | 2 | 5176 | 14 | 2 | 5188 | 6 | 2 | 5192 | 8 | 4 | 5204 | 25 | 2 |
| 5208 | 2 | 8 | 5224 | 9 | 2 | 5236 | 3 | 8 | 5240 | 9 | 4 | 5252 | 6 | 4 | 5268 | 5 | 4 |
| 5272 | 5 | 2 | 5284 | 12 | 2 | 5288 | 21 | 2 | 5304 | 5 | 8 | 5316 | 9 | 4 | 5320 | 3 | 8 |
| 5332 | 5 | 4 | 5336 | 8 | 4 | 5348 | 6 | 4 | 5352 | 5 | 4 | 5368 | 5 | 4 | 5380 | 4 | 4 |
| 5384 | 14 | 2 | 5396 | 14 | 2 | 5412 | 2 | 8 | 5416 | 11 | 2 | 5428 | 4 | 4 | 5432 | 6 | 4 |
| 5444 | 30 | 2 | 5448 | 6 | 4 | 5460 | 1 | 16 | 5464 | 9 | 2 | 5480 | 11 | 4 | 5492 | 9 | 2 |
| 5496 | 7 | 4 | 5512 | 5 | 4 | 5524 | 13 | 2 | 5528 | 19 | 2 | 5540 | 12 | 4 | 5556 | 7 | 4 |
| 5560 | 5 | 4 | 5572 | 4 | 4 | 5576 | 12 | 4 | 5588 | 6 | 4 | 5592 | 5 | 4 | 5604 | 7 | 4 |
| 5608 | 7 | 2 | 5620 | 6 | 4 | 5624 | 11 | 4 | 5636 | 18 | 2 | 5640 | 4 | 8 | 5656 | 7 | 4 |
| 5668 | 4 | 4 | 5672 | 17 | 2 | 5704 | 8 | 4 | 5716 | 11 | 2 | 5720 | 4 | 8 | 5732 | 18 | 2 |
| 5736 | 8 | 4 | 5748 | 6 | 4 | 5752 | 8 | 2 | 5764 | 7 | 4 | 5768 | 6 | 4 | 5784 | 8 | 4 |
| 5812 | 7 | 2 | 5816 | 30 | 2 | 5828 | 6 | 4 | 5844 | 7 | 4 | 5848 | 4 | 4 | 5860 | 4 | 4 |
| 5864 | 29 | 2 | 5876 | 14 | 2 | 5892 | 7 | 4 | 5896 | 4 | 4 | 5908 | 4 | 4 | 5912 | 15 | 2 |
| 5924 | 26 | 2 | 5928 | 3 | 8 | 5944 | 10 | 2 | 5956 | 10 | 2 | 5960 | 9 | 4 | 5972 | 11 | 2 |
| 5988 | 5 | 4 | 5992 | 4 | 4 | 6004 | 6 | 4 | 6008 | 12 | 2 | 6020 | 5 | 8 | 6024 | 6 | 4 |
| 6036 | 10 | 4 | 6040 | 4 | 4 | 6052 | 4 | 4 | 6056 | 25 | 2 | 6068 | 12 | 4 | 6072 | 3 | 8 |
| 6088 | 10 | 2 | 6104 | 12 | 4 | 6116 | 13 | 4 | 6132 | 3 | 8 | 6136 | 5 | 4 | 6148 | 4 | 4 |
| 6152 | 22 | 2 | 6164 | 9 | 4 | 6168 | 5 | 4 | 6180 | 3 | 8 | 6184 | 17 | 2 | 6196 | 9 | 2 |
| 6212 | 20 | 2 | 6216 | 3 | 8 | 6232 | 3 | 4 | 6244 | 8 | 4 | 6248 | 7 | 4 | 6260 | 7 | 4 |
| 6276 | 7 | 4 | 6280 | 5 | 4 | 6296 | 27 | 2 | 6308 | 7 | 4 | 6312 | 4 | 4 | 6324 | 5 | 8 |
| 6328 | 4 | 4 | 6340 | 6 | 4 | 6344 | 11 | 4 | 6356 | 13 | 4 | 6360 | 4 | 8 | 6376 | 17 | 2 |
| 6388 | 7 | 2 | 6392 | 8 | 4 | 6404 | 28 | 2 | 6420 | 2 | 8 | 6424 | 7 | 4 | 6436 | 14 | 2 |
| 6440 | 4 | 8 | 6452 | 21 | 2 | 6456 | 7 | 4 | 6472 | 6 | 2 | 6484 | 9 | 2 | 6488 | 15 | 2 |

Table 1B

| Disc | H/G | G | Disc | H/G | G | Disc | H/G | G | Disc | H/G | G | Disc | H/G | G | Disc | H/G | G |
|---|---|---|---|---|---|---|---|---|---|---|---|---|---|---|---|---|---|
| 6504 | 5 | 4 | 6520 | 7 | 4 | 6532 | 4 | 4 | 6536 | 16 | 4 | 6548 | 19 | 2 | 6564 | 11 | 4 |
| 6568 | 7 | 2 | 6580 | 2 | 8 | 6584 | 22 | 2 | 6596 | 12 | 4 | 6612 | 2 | 8 | 6616 | 11 | 2 |
| 6628 | 8 | 2 | 6632 | 21 | 2 | 6644 | 12 | 4 | 6648 | 5 | 4 | 6676 | 13 | 2 | 6680 | 7 | 4 |
| 6692 | 8 | 4 | 6708 | 2 | 8 | 6712 | 10 | 2 | 6740 | 13 | 4 | 6744 | 11 | 4 | 6756 | 9 | 4 |
| 6772 | 11 | 2 | 6788 | 14 | 2 | 6792 | 4 | 4 | 6808 | 5 | 4 | 6820 | 2 | 8 | 6824 | 29 | 2 |
| 6836 | 21 | 2 | 6852 | 9 | 4 | 6856 | 10 | 2 | 6868 | 4 | 4 | 6872 | 23 | 2 | 6884 | 26 | 2 |
| 6888 | 3 | 8 | 6904 | 12 | 2 | 6916 | 3 | 8 | 6920 | 9 | 4 | 6932 | 17 | 2 | 6952 | 4 | 4 |
| 6964 | 13 | 2 | 6968 | 11 | 4 | 6980 | 10 | 4 | 6996 | 5 | 8 | 7012 | 10 | 2 | 7016 | 19 | 2 |
| 7028 | 7 | 4 | 7032 | 5 | 4 | 7044 | 5 | 4 | 7048 | 12 | 2 | 7060 | 5 | 4 | 7064 | 25 | 2 |
| 7076 | 16 | 4 | 7080 | 5 | 8 | 7096 | 10 | 2 | 7108 | 12 | 2 | 7112 | 10 | 4 | 7124 | 17 | 4 |
| 7140 | 2 | 16 | 7144 | 5 | 4 | 7156 | 13 | 2 | 7160 | 13 | 4 | 7172 | 9 | 4 | 7176 | 4 | 8 |
| 7188 | 6 | 4 | 7192 | 5 | 4 | 7204 | 14 | 2 | 7208 | 8 | 4 | 7224 | 5 | 8 | 7240 | 5 | 4 |
| 7256 | 23 | 2 | 7268 | 10 | 4 | 7284 | 9 | 4 | 7288 | 8 | 2 | 7304 | 14 | 4 | 7316 | 10 | 4 |
| 7320 | 3 | 8 | 7332 | 3 | 8 | 7336 | 9 | 4 | 7348 | 6 | 4 | 7352 | 14 | 2 | 7364 | 10 | 4 |
| 7368 | 4 | 4 | 7384 | 7 | 4 | 7412 | 9 | 4 | 7428 | 5 | 4 | 7432 | 10 | 2 | 7444 | 19 | 2 |
| 7460 | 12 | 4 | 7464 | 8 | 4 | 7476 | 5 | 8 | 7480 | 2 | 8 | 7492 | 6 | 2 | 7496 | 28 | 2 |
| 7508 | 17 | 2 | 7512 | 6 | 4 | 7528 | 9 | 2 | 7540 | 2 | 8 | 7544 | 16 | 4 | 7556 | 36 | 2 |
| 7572 | 5 | 4 | 7576 | 15 | 2 | 7588 | 4 | 4 | 7592 | 7 | 4 | 7604 | 21 | 2 | 7608 | 9 | 4 |
| 7620 | 3 | 8 | 7624 | 10 | 2 | 7636 | 7 | 4 | 7640 | 14 | 4 | 7652 | 18 | 2 | 7656 | 6 | 8 |
| 7672 | 4 | 4 | 7684 | 10 | 4 | 7716 | 7 | 4 | 7720 | 5 | 4 | 7732 | 9 | 2 | 7736 | 26 | 2 |
| 7748 | 12 | 4 | 7752 | 4 | 8 | 7764 | 6 | 4 | 7768 | 11 | 2 | 7780 | 4 | 4 | 7784 | 17 | 4 |
| 7796 | 35 | 2 | 7816 | 14 | 2 | 7828 | 4 | 4 | 7832 | 8 | 4 | 7844 | 8 | 4 | 7860 | 5 | 8 |
| 7864 | 18 | 2 | 7876 | 5 | 4 | 7880 | 13 | 4 | 7892 | 21 | 2 | 7896 | 4 | 8 | 7908 | 9 | 4 |
| 7912 | 3 | 4 | 7924 | 5 | 4 | 7928 | 12 | 2 | 7940 | 10 | 4 | 7944 | 12 | 4 | 7960 | 6 | 4 |
| 7972 | 12 | 2 | 7976 | 27 | 2 | 7988 | 21 | 2 | 8004 | 6 | 8 | 8008 | 2 | 8 | 8020 | 8 | 4 |
| 8024 | 12 | 4 | 8040 | 4 | 8 | 8052 | 2 | 8 | 8056 | 9 | 4 | 8068 | 6 | 2 | 8072 | 14 | 2 |
| 8084 | 17 | 4 | 8088 | 6 | 4 | 8104 | 17 | 2 | 8116 | 17 | 2 | 8120 | 5 | 8 | 8132 | 7 | 4 |
| 8148 | 3 | 8 | 8152 | 9 | 2 | 8164 | 8 | 4 | 8168 | 25 | 2 | 8180 | 12 | 4 | 8184 | 4 | 8 |
| 8196 | 9 | 4 | 8212 | 9 | 2 | 8216 | 18 | 4 | 8248 | 6 | 2 | 8260 | 3 | 4 | 8264 | 28 | 2 |
| 8276 | 19 | 2 | 8292 | 5 | 4 | 8296 | 5 | 4 | 8308 | 4 | 4 | 8312 | 20 | 2 | 8324 | 30 | 2 |
| 8328 | 6 | 4 | 8340 | 4 | 8 | 8344 | 6 | 4 | 8356 | 22 | 2 | 8360 | 6 | 8 | 8372 | 5 | 8 |
| 8376 | 7 | 4 | 8392 | 12 | 2 | 8404 | 5 | 4 | 8408 | 13 | 2 | 8420 | 12 | 4 | 8436 | 5 | 8 |
| 8440 | 7 | 4 | 8452 | 8 | 2 | 8456 | 14 | 4 | 8468 | 9 | 4 | 8472 | 5 | 4 | 8484 | 5 | 8 |
| 8488 | 9 | 2 | 8504 | 30 | 2 | 8516 | 28 | 2 | 8520 | 4 | 8 | 8536 | 8 | 4 | 8548 | 8 | 2 |
| 8552 | 21 | 2 | 8564 | 39 | 2 | 8580 | 2 | 16 | 8584 | 7 | 4 | 8596 | 7 | 4 | 8612 | 16 | 2 |
| 8616 | 10 | 4 | 8628 | 6 | 4 | 8632 | 5 | 4 | 8644 | 18 | 2 | 8648 | 14 | 4 | 8660 | 7 | 4 |
| 8680 | 2 | 8 | 8692 | 5 | 4 | 8696 | 32 | 2 | 8708 | 10 | 4 | 8724 | 9 | 4 | 8728 | 11 | 2 |
| 8740 | 3 | 8 | 8744 | 21 | 2 | 8756 | 15 | 4 | 8760 | 3 | 8 | 8772 | 3 | 8 | 8776 | 16 | 2 |
| 8792 | 9 | 4 | 8804 | 16 | 4 | 8808 | 5 | 4 | 8824 | 16 | 2 | 8840 | 7 | 8 | 8852 | 21 | 2 |
| 8868 | 7 | 4 | 8872 | 13 | 2 | 8884 | 9 | 2 | 8888 | 13 | 4 | 8904 | 5 | 8 | 8916 | 12 | 4 |
| 8920 | 5 | 4 | 8932 | 2 | 8 | 8936 | 17 | 2 | 8948 | 15 | 2 | 8952 | 7 | 4 | 8968 | 4 | 4 |
| 8980 | 8 | 4 | 8984 | 39 | 2 | 8996 | 10 | 4 | 9012 | 9 | 4 | 9028 | 6 | 4 | 9032 | 14 | 2 |
| 9044 | 9 | 8 | 9048 | 3 | 4 | 9060 | 4 | 8 | 9064 | 9 | 4 | 9076 | 15 | 2 | 9080 | 7 | 4 |
| 9092 | 24 | 2 | 9096 | 14 | 4 | 9112 | 6 | 4 | 9124 | 10 | 2 | 9128 | 10 | 4 | 9140 | 15 | 4 |
| 9156 | 4 | 8 | 9160 | 5 | 4 | 9172 | 7 | 2 | 9176 | 19 | 4 | 9188 | 20 | 2 | 9192 | 8 | 4 |
| 9204 | 6 | 8 | 9208 | 8 | 2 | 9220 | 6 | 4 | 9224 | 32 | 2 | 9236 | 33 | 2 | 9240 | 2 | 16 |
| 9256 | 7 | 4 | 9268 | 6 | 4 | 9272 | 9 | 4 | 9284 | 15 | 4 | 9304 | 11 | 2 | 9316 | 8 | 4 |
| 9320 | 15 | 4 | 9332 | 17 | 2 | 9336 | 11 | 4 | 9348 | 4 | 8 | 9352 | 6 | 4 | 9364 | 15 | 2 |
| 9368 | 21 | 2 | 9380 | 7 | 8 | 9384 | 3 | 8 | 9412 | 4 | 4 | 9416 | 14 | 4 | 9428 | 21 | 2 |
| 9444 | 9 | 4 | 9448 | 13 | 2 | 9460 | 4 | 8 | 9476 | 18 | 4 | 9480 | 3 | 8 | 9492 | 3 | 8 |
| 9496 | 11 | 2 | 9508 | 8 | 2 | 9512 | 15 | 4 | 9524 | 19 | 2 | 9528 | 7 | 4 | 9544 | 16 | 2 |
| 9556 | 17 | 2 | 9560 | 12 | 4 | 9572 | 30 | 2 | 9588 | 4 | 8 | 9592 | 5 | 4 | 9608 | 20 | 2 |
| 9620 | 5 | 8 | 9624 | 15 | 4 | 9636 | 5 | 8 | 9640 | 4 | 4 | 9652 | 6 | 4 | 9656 | 18 | 4 |
| 9668 | 18 | 2 | 9672 | 3 | 8 | 9688 | 5 | 4 | 9704 | 27 | 2 | 9716 | 14 | 4 | 9732 | 5 | 4 |
| 9736 | 16 | 2 | 9748 | 9 | 2 | 9752 | 7 | 4 | 9764 | 38 | 2 | 9768 | 4 | 8 | 9780 | 4 | 8 |

Table 1B

| Disc | H/G | G | Disc | H/G | G | Disc | H/G | G | Disc | H/G | G | Disc | H/G | G | Disc | H/G | G |
|------|-----|---|------|-----|---|------|-----|---|------|-----|---|------|-----|---|------|-----|---|
| 9784 | 14 | 2 | 9796 | 10 | 4 | 9812 | 13 | 4 | 9816 | 8 | 4 | 9832 | 9 | 2 | 9844 | 9 | 4 |
| 9848 | 16 | 2 | 9860 | 4 | 8 | 9876 | 13 | 4 | 9880 | 4 | 8 | 9892 | 10 | 2 | 9896 | 39 | 2 |
| 9908 | 19 | 2 | 9912 | 4 | 8 | 9924 | 13 | 4 | 9928 | 6 | 4 | 9940 | 3 | 8 | 9944 | 14 | 4 |
| 9956 | 11 | 4 | 9960 | 4 | 8 | 9976 | 5 | 4 | 9988 | 7 | 4 | 9992 | 20 | 2 | | | |

# Appendix 2: Tables, Positive Discriminants

The following two tables are of class numbers of forms for fundamental positive discriminants $D$ in the range $0 < D < 10000$. For each discriminant the table has the class number $H$ and the sign $S$ of the fundamental unit.

Table 2A

| Disc | H S | Disc | H S | Disc | H S | Disc | H S | Disc | H S | Disc | H S | Disc | H S | Disc | H S |
|---|---|---|---|---|---|---|---|---|---|---|---|---|---|---|---|
| 5 | 1- | 13 | 1- | 17 | 1- | 21 | 2+ | 29 | 1- | 33 | 2+ | 37 | 1- | 41 | 1- |
| 53 | 1- | 57 | 2+ | 61 | 1- | 65 | 2- | 69 | 2+ | 73 | 1- | 77 | 2+ | 85 | 2- |
| 89 | 1- | 93 | 2+ | 97 | 1- | 101 | 1- | 105 | 4+ | 109 | 1- | 113 | 1- | 129 | 2+ |
| 133 | 2+ | 137 | 1- | 141 | 2+ | 145 | 4- | 149 | 1- | 157 | 1- | 161 | 2+ | 165 | 4+ |
| 173 | 1- | 177 | 2+ | 181 | 1- | 185 | 2- | 193 | 1- | 197 | 1- | 201 | 2+ | 205 | 4+ |
| 209 | 2+ | 213 | 2+ | 217 | 2+ | 221 | 4+ | 229 | 3- | 233 | 1- | 237 | 2+ | 241 | 1- |
| 249 | 2+ | 253 | 2+ | 257 | 3- | 265 | 2- | 269 | 1- | 273 | 4+ | 277 | 1- | 281 | 1- |
| 285 | 4+ | 293 | 1- | 301 | 1- | 305 | 2+ | 309 | 4+ | 313 | 1- | 317 | 1- | 321 | 6+ |
| 329 | 2+ | 337 | 1- | 341 | 2+ | 345 | 4+ | 349 | 1- | 353 | 1- | 357 | 4+ | 365 | 1- |
| 373 | 1- | 377 | 4+ | 381 | 2+ | 385 | 4+ | 389 | 1- | 393 | 2+ | 397 | 1- | 401 | 5- |
| 409 | 1- | 413 | 2+ | 417 | 2+ | 421 | 1- | 429 | 4+ | 433 | 1- | 437 | 2+ | 445 | 4- |
| 449 | 1- | 453 | 2+ | 457 | 1- | 461 | 1- | 465 | 4+ | 469 | 6+ | 473 | 6+ | 481 | 2- |
| 485 | 2- | 489 | 2+ | 493 | 2- | 497 | 2+ | 501 | 2+ | 505 | 8+ | 509 | 1- | 517 | 2+ |
| 521 | 1- | 533 | 2- | 537 | 2+ | 541 | 1- | 545 | 4+ | 553 | 2+ | 557 | 1- | 561 | 4+ |
| 565 | 2- | 569 | 1- | 573 | 2+ | 577 | 7- | 581 | 2+ | 589 | 2+ | 593 | 1- | 597 | 2+ |
| 601 | 1- | 609 | 4+ | 613 | 1- | 617 | 1- | 629 | 2- | 633 | 2+ | 641 | 1- | 645 | 4+ |
| 649 | 2+ | 653 | 1- | 661 | 1- | 665 | 4+ | 669 | 2+ | 673 | 1- | 677 | 1- | 681 | 2+ |
| 685 | 2- | 689 | 8+ | 697 | 6- | 701 | 1- | 705 | 4+ | 709 | 1- | 713 | 2+ | 717 | 2+ |
| 721 | 2+ | 733 | 3- | 737 | 2+ | 741 | 4+ | 745 | 4+ | 749 | 2+ | 753 | 2+ | 757 | 1- |
| 761 | 3- | 769 | 1- | 773 | 1- | 777 | 8+ | 781 | 2+ | 785 | 6- | 789 | 2+ | 793 | 8+ |
| 797 | 1- | 805 | 4+ | 809 | 1- | 813 | 2+ | 817 | 10+ | 821 | 1- | 829 | 1- | 849 | 2+ |
| 853 | 1- | 857 | 1- | 861 | 4+ | 865 | 2- | 869 | 2+ | 877 | 1- | 881 | 1- | 885 | 4+ |
| 889 | 2+ | 893 | 2+ | 897 | 8+ | 901 | 4- | 905 | 8+ | 913 | 2+ | 917 | 2+ | 921 | 2+ |
| 929 | 1- | 933 | 2+ | 937 | 1- | 941 | 1- | 949 | 2- | 953 | 1- | 957 | 4+ | 965 | 2- |
| 969 | 4+ | 973 | 2+ | 977 | 1- | 985 | 6- | 989 | 2+ | 993 | 6+ | 997 | 1- | 1001 | 4+ |
| 1005 | 4+ | 1009 | 7- | 1013 | 1- | 1021 | 1- | 1033 | 1- | 1037 | 2- | 1041 | 2+ | 1045 | 8+ |
| 1049 | 1- | 1057 | 2+ | 1061 | 1- | 1065 | 4+ | 1069 | 1- | 1073 | 2- | 1077 | 2+ | 1081 | 2+ |
| 1085 | 4+ | 1093 | 5- | 1097 | 1- | 1101 | 6+ | 1105 | 4- | 1109 | 1- | 1113 | 4+ | 1117 | 1- |
| 1121 | 2+ | 1129 | 9- | 1133 | 2+ | 1137 | 2+ | 1141 | 2+ | 1145 | 4- | 1149 | 2+ | 1153 | 1- |
| 1157 | 2- | 1165 | 2- | 1169 | 1- | 1173 | 2+ | 1177 | 2+ | 1181 | 1- | 1185 | 4+ | 1189 | 2- |
| 1193 | 1- | 1201 | 1- | 1205 | 4+ | 1209 | 4+ | 1213 | 1- | 1217 | 1- | 1221 | 8+ | 1229 | 3- |
| 1237 | 1- | 1241 | 2- | 1245 | 4+ | 1249 | 1- | 1253 | 2+ | 1257 | 6+ | 1261 | 2- | 1265 | 4+ |
| 1273 | 2+ | 1277 | 1- | 1281 | 4+ | 1285 | 2- | 1289 | 1- | 1293 | 2+ | 1297 | 11- | 1301 | 1- |
| 1309 | 2+ | 1313 | 4- | 1317 | 2+ | 1321 | 1- | 1329 | 2+ | 1333 | 2+ | 1337 | 2+ | 1345 | 12+ |
| 1349 | 2+ | 1353 | 4+ | 1357 | 2+ | 1361 | 1- | 1365 | 8+ | 1373 | 3- | 1381 | 1- | 1385 | 2- |
| 1389 | 2+ | 1393 | 10+ | 1397 | 2+ | 1401 | 2+ | 1405 | 4+ | 1409 | 1- | 1417 | 2- | 1429 | 3- |
| 1433 | 1- | 1437 | 2+ | 1441 | 2+ | 1453 | 1- | 1457 | 2+ | 1461 | 2+ | 1465 | 2- | 1469 | 4+ |
| 1473 | 2+ | 1477 | 2+ | 1481 | 1- | 1489 | 3- | 1493 | 1- | 1497 | 2+ | 1501 | 2+ | 1505 | 4+ |
| 1509 | 6+ | 1513 | 4+ | 1517 | 4+ | 1529 | 4+ | 1533 | 4+ | 1537 | 4+ | 1541 | 2+ | 1545 | 4+ |
| 1549 | 1- | 1553 | 1- | 1561 | 2+ | 1565 | 2- | 1569 | 2+ | 1577 | 2+ | 1581 | 4+ | 1585 | 2- |
| 1589 | 2+ | 1597 | 1- | 1601 | 5- | 1605 | 4+ | 1609 | 1- | 1613 | 1- | 1621 | 1- | 1633 | 2+ |
| 1637 | 1- | 1641 | 10+ | 1645 | 4+ | 1649 | 2- | 1653 | 4+ | 1657 | 1- | 1661 | 2+ | 1669 | 1- |
| 1673 | 2+ | 1677 | 8+ | 1685 | 2- | 1689 | 2+ | 1693 | 1- | 1697 | 1- | 1705 | 16+ | 1709 | 1- |
| 1713 | 1- | 1717 | 4+ | 1721 | 1- | 1729 | 4- | 1733 | 1- | 1741 | 1- | 1745 | 4+ | 1749 | 4+ |
| 1753 | 1- | 1757 | 2+ | 1761 | 14+ | 1765 | 4- | 1769 | 2- | 1777 | 1- | 1781 | 1- | 1785 | 12+ |
| 1789 | 1- | 1793 | 2+ | 1797 | 2+ | 1801 | 1- | 1817 | 2+ | 1821 | 2+ | 1829 | 2+ | 1833 | 2+ |
| 1837 | 2+ | 1841 | 2+ | 1853 | 2- | 1857 | 2+ | 1861 | 1- | 1865 | 2- | 1869 | 4+ | 1873 | 1- |
| 1877 | 1- | 1885 | 6+ | 1889 | 1- | 1893 | 2+ | 1897 | 10+ | 1901 | 3- | 1905 | 4+ | 1909 | 2+ |
| 1913 | 1- | 1921 | 2- | 1929 | 6+ | 1933 | 1- | 1937 | 5- | 1941 | 2+ | 1945 | 4+ | 1949 | 1- |
| 1957 | 6+ | 1961 | 4+ | 1965 | 4+ | 1969 | 2+ | 1973 | 1- | 1977 | 2+ | 1981 | 2+ | 1985 | 2- |
| 1993 | 1- | 1997 | 1- | 2001 | 4+ | 2005 | 6+ | 2013 | 4+ | 2017 | 1- | 2021 | 6+ | 2029 | 5- |
| 2033 | 2+ | 2037 | 4+ | 2041 | 4+ | 2045 | 2+ | 2049 | 2+ | 2053 | 1- | 2065 | 4+ | 2069 | 1- |
| 2073 | 2+ | 2077 | 2+ | 2081 | 5- | 2085 | 4+ | 2089 | 3- | 2093 | 4+ | 2101 | 6+ | 2105 | 4+ |
| 2109 | 4+ | 2113 | 1- | 2117 | 1- | 2121 | 4+ | 2129 | 1- | 2137 | 1- | 2141 | 1- | 2145 | 6+ |
| 2149 | 2+ | 2153 | 5- | 2157 | 2+ | 2161 | 1- | 2165 | 2- | 2173 | 2- | 2177 | 6+ | 2181 | 2+ |

Table 2A

| Disc | H S | Disc | H S | Disc | H S | Disc | H S | Disc | H S | Disc | H S | Disc | H S | Disc | H S |
|---|---|---|---|---|---|---|---|---|---|---|---|---|---|---|---|
| 2185 | 4+ | 2189 | 2+ | 2193 | 4+ | 2201 | 2+ | 2213 | 1- | 2217 | 2+ | 2221 | 1- | 2229 | 2+ |
| 2233 | 12+ | 2237 | 1- | 2245 | 4+ | 2249 | 4+ | 2253 | 4- | 2257 | 2+ | 2261 | 4+ | 2265 | 4+ |
| 2269 | 1- | 2273 | 1- | 2281 | 1- | 2285 | 2- | 2289 | 8+ | 2293 | 1- | 2297 | 1- | 2301 | 4+ |
| 2305 | 12- | 2309 | 1- | 2317 | 2+ | 2321 | 2+ | 2329 | 4+ | 2333 | 1- | 2337 | 4+ | 2341 | 1- |
| 2345 | 4+ | 2353 | 4+ | 2357 | 1- | 2361 | 2+ | 2365 | 4+ | 2369 | 2+ | 2373 | 4+ | 2377 | 1- |
| 2381 | 1- | 2389 | 1- | 2393 | 1- | 2397 | 4+ | 2405 | 3- | 2409 | 4+ | 2413 | 2+ | 2417 | 1- |
| 2429 | 6+ | 2433 | 2+ | 2437 | 1- | 2441 | 1- | 2445 | 4+ | 2449 | 2+ | 2453 | 2+ | 2461 | 2+ |
| 2465 | 3- | 2469 | 2+ | 2473 | 1- | 2477 | 1- | 2481 | 2+ | 2485 | 4+ | 2489 | 2+ | 2497 | 2+ |
| 2501 | 3- | 2505 | 8+ | 2509 | 2- | 2513 | 2+ | 2517 | 2+ | 2521 | 1- | 2533 | 6+ | 2537 | 2+ |
| 2545 | 4- | 2549 | 1- | 2553 | 4+ | 2557 | 3- | 2561 | 2- | 2569 | 2+ | 2573 | 2+ | 2577 | 2+ |
| 2581 | 2- | 2585 | 4+ | 2589 | 6+ | 2593 | 1- | 2605 | 6- | 2609 | 1- | 2613 | 4+ | 2617 | 1- |
| 2621 | 1- | 2629 | 2+ | 2633 | 1- | 2641 | 2+ | 2649 | 2+ | 2653 | 2+ | 2657 | 1- | 2661 | 2+ |
| 2665 | 4- | 2669 | 4+ | 2677 | 1- | 2681 | 2+ | 2685 | 4+ | 2689 | 1- | 2693 | 1- | 2697 | 4+ |
| 2701 | 2+ | 2705 | 6- | 2713 | 3- | 2717 | 4+ | 2721 | 2+ | 2729 | 1- | 2733 | 2+ | 2737 | 4+ |
| 2741 | 1- | 2749 | 1- | 2753 | 1- | 2757 | 2+ | 2761 | 2+ | 2765 | 4+ | 2769 | 4+ | 2773 | 2+ |
| 2777 | 1- | 2785 | 2- | 2789 | 1- | 2797 | 1- | 2801 | 1- | 2805 | 6+ | 2813 | 1- | 2821 | 4+ |
| 2829 | 4+ | 2833 | 1- | 2837 | 1- | 2841 | 2+ | 2845 | 4+ | 2849 | 8+ | 2857 | 3- | 2861 | 1- |
| 2865 | 4+ | 2869 | 2+ | 2877 | 4+ | 2881 | 2+ | 2885 | 2- | 2893 | 2+ | 2897 | 1- | 2901 | 2+ |
| 2905 | 4+ | 2909 | 1- | 2913 | 10+ | 2917 | 1- | 2921 | 2+ | 2929 | 2- | 2933 | 2+ | 2937 | 4+ |
| 2941 | 2- | 2945 | 4+ | 2949 | 2+ | 2953 | 1- | 2957 | 1- | 2965 | 2+ | 2969 | 1- | 2973 | 2+ |
| 2977 | 2- | 2981 | 6+ | 2985 | 4+ | 2993 | 12+ | 3001 | 1- | 3005 | 4+ | 3009 | 4+ | 3013 | 2+ |
| 3017 | 2+ | 3021 | 12+ | 3029 | 2- | 3037 | 1- | 3041 | 1- | 3045 | 8+ | 3049 | 1- | 3053 | 2+ |
| 3057 | 2+ | 3061 | 1- | 3065 | 2- | 3073 | 2+ | 3077 | 2- | 3081 | 16+ | 3085 | 2- | 3089 | 1- |
| 3093 | 2+ | 3097 | 2+ | 3101 | 2+ | 3109 | 1- | 3113 | 2+ | 3117 | 2+ | 3121 | 5- | 3129 | 8+ |
| 3133 | 2+ | 3137 | 5- | 3145 | 4- | 3149 | 2+ | 3153 | 2+ | 3157 | 4+ | 3161 | 2- | 3165 | 4+ |
| 3169 | 1- | 3173 | 2+ | 3181 | 5- | 3189 | 2+ | 3193 | 2+ | 3197 | 2+ | 3201 | 16+ | 3205 | 4+ |
| 3209 | 1- | 3217 | 1- | 3221 | 3- | 3229 | 3- | 3233 | 2- | 3237 | 4+ | 3241 | 2+ | 3245 | 4+ |
| 3253 | 5- | 3257 | 1- | 3261 | 6+ | 3265 | 2- | 3269 | 2+ | 3273 | 2+ | 3277 | 2- | 3281 | 5- |
| 3289 | 4+ | 3293 | 2- | 3297 | 4+ | 3301 | 1- | 3305 | 12+ | 3309 | 2+ | 3313 | 1- | 3317 | 2+ |
| 3329 | 1- | 3333 | 4+ | 3337 | 2+ | 3341 | 4- | 3345 | 4+ | 3349 | 1- | 3353 | 2+ | 3361 | 1- |
| 3365 | 2- | 3369 | 2+ | 3373 | 1- | 3377 | 2+ | 3385 | 2- | 3389 | 1- | 3397 | 2+ | 3401 | 2+ |
| 3405 | 4+ | 3409 | 2+ | 3413 | 1- | 3417 | 4+ | 3421 | 2+ | 3433 | 1- | 3437 | 2+ | 3441 | 4+ |
| 3445 | 4- | 3449 | 1- | 3453 | 2+ | 3457 | 1- | 3461 | 1- | 3469 | 1- | 3473 | 2+ | 3477 | 8+ |
| 3485 | 3- | 3489 | 2+ | 3493 | 2+ | 3497 | 4+ | 3505 | 4+ | 3513 | 2+ | 3517 | 1- | 3521 | 2+ |
| 3529 | 1- | 3533 | 1- | 3541 | 1- | 3545 | 4- | 3553 | 4+ | 3557 | 1- | 3561 | 2+ | 3565 | 4+ |
| 3569 | 6+ | 3581 | 1- | 3585 | 16+ | 3589 | 2- | 3593 | 1- | 3597 | 4+ | 3601 | 12- | 3605 | 4+ |
| 3613 | 1- | 3617 | 1- | 3621 | 4+ | 3629 | 2+ | 3633 | 4+ | 3637 | 1- | 3641 | 2+ | 3649 | 2- |
| 3653 | 2- | 3657 | 4+ | 3661 | 2+ | 3665 | 2- | 3669 | 2+ | 3673 | 1- | 3677 | 1- | 3685 | 4+ |
| 3689 | 4+ | 3693 | 2+ | 3697 | 1- | 3701 | 1- | 3705 | 8+ | 3709 | 1- | 3713 | 2+ | 3729 | 4+ |
| 3733 | 1- | 3737 | 4+ | 3741 | 4+ | 3745 | 4+ | 3749 | 2+ | 3761 | 1- | 3765 | 4+ | 3769 | 1- |
| 3777 | 2+ | 3781 | 2+ | 3785 | 2- | 3793 | 1- | 3797 | 1- | 3801 | 4+ | 3805 | 4+ | 3809 | 2- |
| 3813 | 4+ | 3817 | 2+ | 3821 | 1- | 3829 | 2+ | 3833 | 1- | 3837 | 2+ | 3841 | 2+ | 3845 | 4- |
| 3849 | 3- | 3853 | 1- | 3857 | 4+ | 3865 | 2- | 3869 | 3- | 3873 | 6+ | 3877 | 3- | 3881 | 1- |
| 3885 | 8+ | 3889 | 3- | 3893 | 4+ | 3901 | 4+ | 3905 | 4+ | 3909 | 2+ | 3913 | 4+ | 3917 | 1- |
| 3921 | 2+ | 3929 | 1- | 3937 | 2+ | 3941 | 6+ | 3945 | 4+ | 3949 | 2+ | 3953 | 2+ | 3957 | 6+ |
| 3961 | 2- | 3965 | 6+ | 3973 | 4- | 3977 | 2- | 3981 | 2+ | 3985 | 2- | 3989 | 1- | 3997 | 6+ |
| 4001 | 3- | 4009 | 18+ | 4013 | 1- | 4017 | 8+ | 4021 | 1- | 4029 | 4+ | 4033 | 2- | 4037 | 2+ |
| 4045 | 2- | 4049 | 1- | 4053 | 8+ | 4057 | 1- | 4061 | 2+ | 4065 | 12+ | 4069 | 4+ | 4073 | 1- |
| 4081 | 8+ | 4085 | 4+ | 4089 | 4+ | 4093 | 1- | 4097 | 4- | 4101 | 2+ | 4105 | 4+ | 4109 | 2+ |
| 4117 | 2+ | 4121 | 2- | 4129 | 1- | 4133 | 1- | 4137 | 4+ | 4141 | 2- | 4145 | 4+ | 4153 | 1- |
| 4157 | 1- | 4161 | 14+ | 4169 | 2+ | 4173 | 8+ | 4177 | 1- | 4181 | 2- | 4189 | 2+ | 4193 | 6+ |
| 4197 | 2+ | 4201 | 1- | 4209 | 4+ | 4213 | 2+ | 4217 | 1- | 4229 | 5- | 4233 | 4+ | 4237 | 2+ |
| 4241 | 1- | 4245 | 4+ | 4249 | 2+ | 4253 | 1- | 4261 | 1- | 4265 | 2- | 4269 | 2+ | 4273 | 1- |
| 4277 | 4+ | 4281 | 6+ | 4285 | 2- | 4289 | 1- | 4297 | 1- | 4301 | 4+ | 4305 | 14+ | 4309 | 2+ |
| 4313 | 2+ | 4317 | 2+ | 4321 | 16+ | 4333 | 2+ | 4337 | 1- | 4341 | 2+ | 4345 | 16+ | 4349 | 1- |

Table 2A

| Disc | H S | Disc | H S | Disc | H S | Disc | H S | Disc | H S | Disc | H S | Disc | H S | Disc | H S |
|---|---|---|---|---|---|---|---|---|---|---|---|---|---|---|---|
| 4353 | 2+ | 4357 | 3- | 4369 | 4+ | 4373 | 1- | 4377 | 2+ | 4381 | 4+ | 4385 | 2- | 4389 | 6+ |
| 4393 | 2+ | 4397 | 1- | 4405 | 4+ | 4409 | 5- | 4413 | 2+ | 4417 | 2+ | 4421 | 1- | 4429 | 2+ |
| 4433 | 4+ | 4441 | 5- | 4445 | 4+ | 4449 | 2+ | 4453 | 4+ | 4457 | 1- | 4461 | 2+ | 4465 | 4+ |
| 4469 | 1- | 4481 | 3- | 4485 | 8+ | 4493 | 1- | 4497 | 2+ | 4501 | 2+ | 4505 | 4- | 4513 | 1- |
| 4517 | 1- | 4521 | 4+ | 4529 | 2+ | 4533 | 2+ | 4537 | 2- | 4541 | 2+ | 4549 | 1- | 4553 | 2- |
| 4561 | 1- | 4565 | 4+ | 4569 | 2+ | 4573 | 2- | 4577 | 2+ | 4585 | 4+ | 4589 | 1- | 4593 | 2+ |
| 4597 | 1- | 4601 | 2+ | 4605 | 4+ | 4609 | 2+ | 4613 | 2+ | 4621 | 1- | 4629 | 2+ | 4633 | 4+ |
| 4637 | 1- | 4641 | 14+ | 4645 | 4+ | 4649 | 3- | 4657 | 1- | 4661 | 2+ | 4665 | 4+ | 4669 | 12+ |
| 4673 | 1- | 4677 | 2+ | 4681 | 2+ | 4685 | 2- | 4697 | 4+ | 4701 | 2+ | 4705 | 8+ | 4709 | 2- |
| 4713 | 2+ | 4717 | 4+ | 4721 | 1- | 4729 | 3- | 4733 | 1- | 4737 | 2+ | 4741 | 2+ | 4745 | 2- |
| 4749 | 6+ | 4757 | 6+ | 4765 | 4+ | 4769 | 2+ | 4773 | 2+ | 4777 | 2- | 4781 | 2+ | 4785 | 8+ |
| 4789 | 1- | 4793 | 1- | 4801 | 1- | 4809 | 4+ | 4813 | 1- | 4817 | 1- | 4821 | 2+ | 4829 | 2+ |
| 4837 | 2+ | 4841 | 6+ | 4845 | 8+ | 4849 | 4+ | 4853 | 2+ | 4857 | 6+ | 4861 | 1- | 4865 | 8+ |
| 4873 | 2+ | 4877 | 1- | 4881 | 2+ | 4885 | 2- | 4889 | 3- | 4893 | 4+ | 4897 | 2+ | 4909 | 1- |
| 4917 | 4+ | 4921 | 4+ | 4929 | 4+ | 4933 | 3- | 4937 | 2- | 4945 | 4+ | 4953 | 8+ | 4957 | 1- |
| 4965 | 4+ | 4969 | 1- | 4973 | 1- | 4981 | 8+ | 4985 | 2- | 4989 | 2+ | 4993 | 1- | 4997 | 2+ |
| 5001 | 2+ | 5005 | 8+ | 5009 | 1- | 5017 | 4+ | 5021 | 1- | 5029 | 2+ | 5033 | 2+ | 5037 | 4+ |
| 5045 | 2- | 5053 | 2+ | 5057 | 4+ | 5061 | 4+ | 5065 | 2- | 5069 | 4+ | 5073 | 12+ | 5077 | 1- |
| 5081 | 3- | 5089 | 6+ | 5093 | 2+ | 5097 | 2+ | 5101 | 1- | 5105 | 6+ | 5109 | 8+ | 5113 | 1- |
| 5117 | 4+ | 5129 | 2+ | 5133 | 4+ | 5137 | 2+ | 5141 | 2+ | 5149 | 2+ | 5153 | 1- | 5161 | 2- |
| 5165 | 2- | 5169 | 2+ | 5173 | 2+ | 5177 | 2+ | 5181 | 4+ | 5185 | 9- | 5189 | 1- | 5197 | 1- |
| 5201 | 2+ | 5205 | 4+ | 5209 | 1- | 5213 | 2- | 5217 | 8+ | 5221 | 2+ | 5233 | 1- | 5237 | 1- |
| 5241 | 6+ | 5245 | 4- | 5249 | 12+ | 5253 | 4+ | 5257 | 2+ | 5261 | 3- | 5269 | 10+ | 5273 | 1- |
| 5277 | 2+ | 5281 | 3- | 5285 | 4+ | 5289 | 4+ | 5293 | 2+ | 5297 | 1- | 5305 | 8- | 5309 | 1- |
| 5313 | 8+ | 5317 | 2- | 5321 | 2+ | 5333 | 1- | 5345 | 4+ | 5349 | 2+ | 5353 | 5- | 5357 | 2+ |
| 5361 | 2+ | 5365 | 3- | 5369 | 12+ | 5377 | 2+ | 5381 | 1- | 5385 | 4+ | 5389 | 2- | 5393 | 2+ |
| 5397 | 4+ | 5401 | 2+ | 5405 | 4+ | 5413 | 1- | 5417 | 5- | 5421 | 4+ | 5429 | 2- | 5433 | 2+ |
| 5437 | 1- | 5441 | 1- | 5449 | 1- | 5453 | 4+ | 5457 | 4+ | 5461 | 2+ | 5465 | 2- | 5469 | 2+ |
| 5473 | 2- | 5477 | 1- | 5485 | 2- | 5489 | 2+ | 5493 | 2+ | 5497 | 6+ | 5501 | 1- | 5505 | 4+ |
| 5509 | 2+ | 5513 | 6+ | 5521 | 7- | 5529 | 8+ | 5533 | 2+ | 5541 | 2+ | 5545 | 4- | 5549 | 2- |
| 5557 | 1- | 5561 | 2+ | 5565 | 6+ | 5569 | 1- | 5573 | 1- | 5581 | 1- | 5585 | 2- | 5593 | 4+ |
| 5597 | 2- | 5601 | 2+ | 5605 | 8+ | 5609 | 2+ | 5613 | 6+ | 5617 | 2- | 5621 | 12+ | 5629 | 4- |
| 5633 | 2+ | 5637 | 6+ | 5641 | 1- | 5645 | 4+ | 5649 | 4+ | 5653 | 1- | 5657 | 1- | 5665 | 4+ |
| 5669 | 1- | 5673 | 4+ | 5677 | 2+ | 5681 | 4+ | 5685 | 8+ | 5689 | 1- | 5693 | 1- | 5701 | 1- |
| 5705 | 4+ | 5709 | 4+ | 5713 | 6- | 5717 | 1- | 5721 | 2+ | 5729 | 2- | 5737 | 1- | 5741 | 1- |
| 5745 | 4+ | 5749 | 1- | 5753 | 2+ | 5757 | 4+ | 5761 | 2+ | 5765 | 2- | 5773 | 2+ | 5777 | 6- |
| 5781 | 4+ | 5785 | 3- | 5789 | 2+ | 5793 | 2+ | 5797 | 4+ | 5801 | 1- | 5809 | 4+ | 5813 | 1- |
| 5817 | 8+ | 5821 | 3- | 5829 | 4+ | 5833 | 2+ | 5837 | 2- | 5845 | 4+ | 5849 | 1- | 5853 | 6+ |
| 5857 | 1- | 5861 | 1- | 5865 | 14+ | 5869 | 1- | 5873 | 2+ | 5881 | 4+ | 5885 | 4+ | 5889 | 4+ |
| 5893 | 2+ | 5897 | 1- | 5901 | 8+ | 5905 | 4+ | 5909 | 2+ | 5917 | 4+ | 5921 | 2+ | 5933 | 2- |
| 5937 | 2+ | 5941 | 2- | 5945 | 8+ | 5953 | 1- | 5957 | 4+ | 5961 | 2+ | 5965 | 2- | 5969 | 2+ |
| 5973 | 4+ | 5977 | 2+ | 5981 | 1- | 5989 | 4+ | 5993 | 2- | 5997 | 2+ | 6001 | 8+ | 6005 | 4+ |
| 6009 | 2+ | 6013 | 2+ | 6017 | 2+ | 6029 | 1- | 6033 | 2+ | 6037 | 1- | 6041 | 2+ | 6045 | 6+ |
| 6049 | 2+ | 6053 | 1- | 6061 | 4+ | 6065 | 2- | 6073 | 1- | 6077 | 2+ | 6081 | 6+ | 6085 | 4- |
| 6089 | 1- | 6097 | 14+ | 6101 | 1- | 6105 | 8+ | 6109 | 1- | 6113 | 3- | 6117 | 2+ | 6121 | 1- |
| 6133 | 3- | 6141 | 4+ | 6145 | 4- | 6149 | 4+ | 6153 | 12+ | 6157 | 2+ | 6161 | 2- | 6169 | 2+ |
| 6173 | 2- | 6177 | 4+ | 6181 | 2+ | 6185 | 2+ | 6189 | 2+ | 6193 | 4+ | 6197 | 1- | 6205 | 3- |
| 6209 | 6+ | 6213 | 4+ | 6217 | 1- | 6221 | 1- | 6229 | 1- | 6233 | 2+ | 6245 | 4- | 6249 | 2+ |
| 6257 | 1- | 6261 | 2+ | 6265 | 4+ | 6269 | 1- | 6277 | 1- | 6281 | 2+ | 6285 | 4+ | 6289 | 6+ |
| 6293 | 4+ | 6297 | 2+ | 6301 | 1- | 6305 | 4- | 6313 | 2+ | 6317 | 1- | 6329 | 1- | 6333 | 2+ |
| 6337 | 1- | 6341 | 2+ | 6349 | 2+ | 6353 | 1- | 6357 | 4+ | 6361 | 1- | 6365 | 4+ | 6369 | 4+ |
| 6373 | 1- | 6377 | 2+ | 6385 | 2- | 6389 | 1- | 6393 | 2+ | 6397 | 1- | 6401 | 3- | 6405 | 8+ |
| 6409 | 6- | 6421 | 1- | 6429 | 2+ | 6433 | 2+ | 6437 | 1- | 6441 | 4+ | 6445 | 4- | 6449 | 1- |
| 6457 | 2+ | 6461 | 4+ | 6465 | 4+ | 6469 | 1- | 6473 | 1- | 6477 | 4+ | 6481 | 5- | 6485 | 2- |
| 6493 | 2+ | 6497 | 2+ | 6501 | 4+ | 6505 | 4+ | 6509 | 2+ | 6513 | 4+ | 6521 | 1- | 6529 | 1- |

Table 2A

| Disc | H S | Disc | H S | Disc | H S | Disc | H S | Disc | H S | Disc | H S | Disc | H S | Disc | H S | Disc | H S |
|---|---|---|---|---|---|---|---|---|---|---|---|---|---|---|---|---|---|
| 6533 | 2+ | 6537 | 2+ | 6541 | 2+ | 6545 | 8+ | 6549 | 6+ | 6553 | 1- | 6557 | 2+ | 6565 | 5- | | |
| 6569 | 1- | 6573 | 4+ | 6577 | 1- | 6581 | 1- | 6585 | 4+ | 6589 | 2+ | 6593 | 2+ | 6601 | 12+ | | |
| 6605 | 2+ | 6609 | 2+ | 6613 | 4+ | 6617 | 2- | 6621 | 2+ | 6629 | 2+ | 6637 | 1- | 6641 | 2- | | |
| 6645 | 4+ | 6649 | 4- | 6653 | 1- | 6657 | 4+ | 6661 | 1- | 6665 | 4+ | 6673 | 1- | 6677 | 2+ | | |
| 6681 | 12+ | 6685 | 8+ | 6689 | 1- | 6693 | 2+ | 6697 | 2+ | 6701 | 1- | 6709 | 1- | 6717 | 2+ | | |
| 6721 | 4+ | 6729 | 2+ | 6733 | 1- | 6737 | 1- | 6745 | 8+ | 6749 | 2- | 6753 | 2+ | 6757 | 6- | | |
| 6761 | 1- | 6765 | 8+ | 6769 | 2+ | 6773 | 4+ | 6781 | 1- | 6785 | 4+ | 6789 | 4+ | 6793 | 1- | | |
| 6797 | 2+ | 6801 | 2+ | 6805 | 8+ | 6809 | 6+ | 6817 | 2- | 6821 | 2+ | 6829 | 1- | 6833 | 1- | | |
| 6837 | 4+ | 6841 | 1- | 6853 | 4+ | 6857 | 1- | 6861 | 2- | 6865 | 2- | 6869 | 1- | 6873 | 4+ | | |
| 6881 | 2+ | 6893 | 4- | 6901 | 2+ | 6905 | 4+ | 6913 | 2+ | 6917 | 1- | 6933 | 2+ | 6937 | 2+ | | |
| 6941 | 2+ | 6945 | 16+ | 6949 | 5- | 6953 | 6- | 6961 | 1- | 6965 | 4+ | 6969 | 4+ | 6973 | 2+ | | |
| 6977 | 1- | 6981 | 4+ | 6985 | 4+ | 6989 | 2- | 6997 | 3- | 7001 | 1- | 7005 | 4+ | 7009 | 2+ | | |
| 7013 | 1- | 7017 | 2+ | 7021 | 4+ | 7033 | 2- | 7037 | 2+ | 7041 | 2+ | 7045 | 2- | 7049 | 8+ | | |
| 7053 | 2+ | 7057 | 9- | 7061 | 2+ | 7069 | 1- | 7073 | 2+ | 7077 | 4+ | 7081 | 4+ | 7085 | 4- | | |
| 7089 | 4+ | 7093 | 2- | 7097 | 2+ | 7109 | 1- | 7113 | 2+ | 7117 | 2+ | 7121 | 1- | 7129 | 1- | | |
| 7133 | 2+ | 7141 | 2- | 7145 | 2+ | 7149 | 2+ | 7153 | 2+ | 7157 | 2+ | 7161 | 8+ | 7165 | 2- | | |
| 7169 | 2+ | 7177 | 1- | 7181 | 2+ | 7185 | 4+ | 7189 | 4+ | 7193 | 1- | 7197 | 2+ | 7201 | 2+ | | |
| 7205 | 8+ | 7213 | 1- | 7217 | 2+ | 7221 | 12+ | 7229 | 1- | 7233 | 2+ | 7237 | 1- | 7241 | 2- | | |
| 7249 | 6+ | 7253 | 1- | 7257 | 4+ | 7261 | 2- | 7265 | 2- | 7269 | 2+ | 7273 | 6+ | 7277 | 2+ | | |
| 7285 | 2+ | 7289 | 6+ | 7293 | 8+ | 7297 | 1- | 7305 | 4+ | 7309 | 1- | 7313 | 2+ | 7321 | 1- | | |
| 7329 | 4+ | 7333 | 1- | 7337 | 4+ | 7341 | 2+ | 7345 | 4- | 7349 | 1- | 7357 | 2+ | 7361 | 2+ | | |
| 7365 | 4+ | 7369 | 1- | 7373 | 1- | 7377 | 2+ | 7385 | 2+ | 7393 | 1- | 7397 | 4- | 7401 | 2+ | | |
| 7405 | 4+ | 7409 | 2+ | 7413 | 4+ | 7417 | 1- | 7421 | 1- | 7429 | 4+ | 7433 | 1- | 7437 | 4+ | | |
| 7441 | 6+ | 7445 | 4+ | 7449 | 4+ | 7453 | 6+ | 7457 | 1- | 7465 | 10- | 7469 | 4+ | 7473 | 8+ | | |
| 7477 | 1- | 7481 | 1- | 7485 | 4+ | 7489 | 1- | 7493 | 2+ | 7501 | 2- | 7505 | 8+ | 7509 | 2+ | | |
| 7513 | 10+ | 7517 | 1- | 7521 | 4+ | 7529 | 1- | 7537 | 3- | 7541 | 3- | 7545 | 4+ | 7549 | 1- | | |
| 7553 | 4+ | 7557 | 8+ | 7561 | 8+ | 7565 | 1- | 7573 | 6+ | 7577 | 1- | 7585 | 6- | 7589 | 1- | | |
| 7593 | 2+ | 7597 | 2+ | 7601 | 6+ | 7609 | 2+ | 7613 | 2+ | 7617 | 2+ | 7621 | 1- | 7629 | 2+ | | |
| 7633 | 2- | 7637 | 2+ | 7645 | 4+ | 7649 | 1- | 7653 | 2+ | 7657 | 4+ | 7661 | 2+ | 7665 | 20+ | | |
| 7669 | 1- | 7673 | 3- | 7681 | 1- | 7685 | 6+ | 7689 | 4+ | 7697 | 2+ | 7701 | 4+ | 7705 | 12+ | | |
| 7709 | 2- | 7717 | 1- | 7721 | 6+ | 7729 | 2+ | 7733 | 4+ | 7737 | 2+ | 7741 | 1- | 7745 | 6- | | |
| 7753 | 3- | 7757 | 1- | 7761 | 8+ | 7765 | 2- | 7769 | 4+ | 7773 | 2+ | 7777 | 4+ | 7781 | 2+ | | |
| 7789 | 1- | 7793 | 1- | 7797 | 4+ | 7801 | 2- | 7805 | 4+ | 7809 | 4+ | 7813 | 2+ | 7817 | 1- | | |
| 7829 | 1- | 7833 | 16+ | 7837 | 4- | 7841 | 1- | 7845 | 4+ | 7849 | 2+ | 7853 | 1- | 7861 | 10+ | | |
| 7869 | 2+ | 7873 | 5- | 7877 | 1- | 7881 | 8+ | 7885 | 4+ | 7897 | 2- | 7901 | 2- | 7905 | 8+ | | |
| 7909 | 2+ | 7913 | 2- | 7917 | 1- | 7933 | 8+ | 7937 | 1- | 7941 | 2+ | 7945 | 4+ | 7949 | 1- | | |
| 7953 | 4+ | 7957 | 2+ | 7961 | 2+ | 7969 | 2- | 7973 | 4+ | 7977 | 2+ | 7981 | 2+ | 7985 | 2- | | |
| 7989 | 2+ | 7993 | 1- | 7997 | 2+ | 8005 | 4- | 8009 | 1- | 8013 | 2+ | 8017 | 3- | 8021 | 2- | | |
| 8029 | 4+ | 8033 | 4+ | 8041 | 4+ | 8045 | 4+ | 8049 | 6+ | 8053 | 1- | 8057 | 6+ | 8061 | 2+ | | |
| 8065 | 2- | 8069 | 3- | 8077 | 2- | 8081 | 1- | 8089 | 1- | 8093 | 1- | 8097 | 6+ | 8101 | 3- | | |
| 8105 | 8+ | 8113 | 8+ | 8117 | 1- | 8121 | 6+ | 8129 | 2+ | 8133 | 2+ | 8137 | 2+ | 8141 | 2+ | | |
| 8149 | 8+ | 8153 | 2+ | 8157 | 2+ | 8161 | 1- | 8165 | 4+ | 8169 | 4+ | 8173 | 6+ | 8177 | 3- | | |
| 8185 | 6- | 8189 | 2+ | 8193 | 2+ | 8197 | 2+ | 8201 | 2+ | 8205 | 4+ | 8209 | 1- | 8213 | 2+ | | |
| 8221 | 1- | 8229 | 4+ | 8233 | 1- | 8237 | 1- | 8241 | 4+ | 8245 | 2- | 8249 | 1- | 8257 | 2+ | | |
| 8261 | 2+ | 8265 | 8+ | 8269 | 1- | 8273 | 1- | 8277 | 4+ | 8285 | 2- | 8293 | 1- | 8297 | 1- | | |
| 8301 | 2+ | 8305 | 8+ | 8309 | 6+ | 8313 | 4+ | 8317 | 1- | 8321 | 7- | 8329 | 1- | 8333 | 2+ | | |
| 8337 | 4+ | 8341 | 2+ | 8345 | 4- | 8353 | 1- | 8357 | 2+ | 8365 | 4+ | 8369 | 1- | 8373 | 6+ | | |
| 8377 | 1- | 8385 | 8+ | 8389 | 1- | 8393 | 4+ | 8401 | 2+ | 8409 | 2+ | 8413 | 2+ | 8417 | 2+ | | |
| 8421 | 4+ | 8429 | 1- | 8437 | 4+ | 8441 | 14+ | 8445 | 4+ | 8449 | 4+ | 8453 | 2+ | 8457 | 2+ | | |
| 8461 | 1- | 8465 | 4- | 8473 | 8+ | 8481 | 4+ | 8485 | 2- | 8489 | 4+ | 8493 | 4+ | 8497 | 2- | | |
| 8501 | 1- | 8509 | 2+ | 8513 | 1- | 8517 | 4+ | 8521 | 1- | 8529 | 2+ | 8533 | 4+ | 8537 | 1- | | |
| 8545 | 6- | 8549 | 2+ | 8553 | 2+ | 8557 | 2+ | 8561 | 2+ | 8565 | 4+ | 8569 | 20+ | 8573 | 1- | | |
| 8581 | 3- | 8585 | 8+ | 8589 | 4+ | 8593 | 1- | 8597 | 3- | 8601 | 8+ | 8605 | 6+ | 8609 | 1- | | |
| 8617 | 2+ | 8621 | 2+ | 8629 | 1- | 8633 | 4+ | 8637 | 2+ | 8641 | 1- | 8645 | 8+ | 8653 | 6- | | |
| 8657 | 2+ | 8661 | 2+ | 8665 | 2- | 8669 | 1- | 8677 | 1- | 8681 | 1- | 8689 | 5- | 8693 | 1- | | |

240

APPENDIX 2

Table 2A

| Disc | H S | Disc | H S | Disc | H S | Disc | H S | Disc | H S | Disc | H S | Disc | H S | Disc | H S |
|---|---|---|---|---|---|---|---|---|---|---|---|---|---|---|---|
| 8697 | 4+ | 8701 | 4+ | 8705 | 8+ | 8709 | 2+ | 8713 | 3- | 8717 | 2+ | 8729 | 4+ | 8733 | 4+ |
| 8737 | 1- | 8741 | 1- | 8745 | 12+ | 8749 | 4+ | 8753 | 1- | 8761 | 9- | 8765 | 2- | 8769 | 12+ |
| 8773 | 2+ | 8777 | 2+ | 8781 | 2+ | 8785 | 4+ | 8789 | 8+ | 8797 | 2+ | 8801 | 8+ | 8805 | 4+ |
| 8809 | 2+ | 8813 | 2+ | 8817 | 2+ | 8821 | 1- | 8837 | 1- | 8841 | 8+ | 8845 | 4+ | 8849 | 1- |
| 8853 | 4+ | 8857 | 2- | 8861 | 1- | 8873 | 2+ | 8877 | 4+ | 8881 | 2+ | 8885 | 2- | 8889 | 2+ |
| 8893 | 1- | 8897 | 1- | 8905 | 4+ | 8909 | 4- | 8913 | 2+ | 8917 | 2- | 8921 | 2+ | 8929 | 1- |
| 8933 | 1- | 8941 | 1- | 8945 | 2- | 8949 | 4+ | 8953 | 4+ | 8961 | 2+ | 8965 | 4+ | 8969 | 1- |
| 8977 | 2+ | 8981 | 2+ | 8985 | 4+ | 8989 | 1- | 8997 | 2+ | 9001 | 1- | 9005 | 4+ | 9013 | 1- |
| 9017 | 2+ | 9021 | 8+ | 9029 | 1- | 9033 | 2+ | 9037 | 2+ | 9041 | 1- | 9049 | 5- | 9053 | 2+ |
| 9057 | 2+ | 9061 | 4- | 9069 | 2+ | 9073 | 6+ | 9077 | 2+ | 9085 | 4+ | 9089 | 4- | 9093 | 4+ |
| 9097 | 2+ | 9101 | 2+ | 9105 | 4+ | 9109 | 1- | 9113 | 4+ | 9121 | 2+ | 9129 | 4+ | 9133 | 3- |
| 9137 | 1- | 9141 | 4+ | 9145 | 8+ | 9149 | 2+ | 9157 | 1- | 9161 | 1- | 9165 | 8+ | 9169 | 2- |
| 9173 | 1- | 9177 | 8+ | 9181 | 3- | 9185 | 4+ | 9193 | 2- | 9197 | 2- | 9201 | 2+ | 9205 | 4+ |
| 9209 | 1- | 9213 | 8+ | 9217 | 4- | 9221 | 1- | 9229 | 2+ | 9233 | 2+ | 9237 | 2+ | 9241 | 1- |
| 9249 | 2+ | 9253 | 2+ | 9257 | 1- | 9265 | 4- | 9269 | 4+ | 9273 | 4+ | 9277 | 1- | 9281 | 3- |
| 9285 | 4+ | 9289 | 6+ | 9293 | 3- | 9301 | 6+ | 9305 | 6- | 9309 | 4+ | 9313 | 2+ | 9321 | 8+ |
| 9329 | 2+ | 9337 | 1- | 9341 | 1- | 9345 | 12+ | 9349 | 1- | 9353 | 2+ | 9357 | 2+ | 9361 | 4+ |
| 9365 | 2- | 9373 | 4+ | 9377 | 1- | 9381 | 4+ | 9385 | 2- | 9389 | 2- | 9393 | 4+ | 9397 | 1- |
| 9401 | 4+ | 9413 | 1- | 9417 | 4+ | 9421 | 1- | 9429 | 4+ | 9433 | 1- | 9437 | 1- | 9445 | 2+ |
| 9449 | 2+ | 9453 | 4+ | 9461 | 4+ | 9465 | 4+ | 9469 | 4+ | 9473 | 4+ | 9481 | 2+ | 9485 | 4+ |
| 9489 | 2+ | 9493 | 2+ | 9497 | 1- | 9501 | 2+ | 9505 | 8+ | 9509 | 2- | 9517 | 2+ | 9521 | 1- |
| 9529 | 2- | 9533 | 1- | 9541 | 2+ | 9545 | 4+ | 9553 | 6- | 9557 | 2+ | 9561 | 2+ | 9565 | 2- |
| 9569 | 2+ | 9573 | 2+ | 9577 | 2- | 9581 | 4+ | 9589 | 2+ | 9593 | 1- | 9597 | 4+ | 9601 | 1- |
| 9605 | 3- | 9609 | 2+ | 9613 | 2+ | 9617 | 2+ | 9629 | 1- | 9637 | 2+ | 9641 | 2+ | 9645 | 4+ |
| 9649 | 1- | 9661 | 1- | 9665 | 2- | 9669 | 8+ | 9673 | 4- | 9677 | 1- | 9681 | 4+ | 9685 | 3- |
| 9689 | 1- | 9697 | 1- | 9701 | 2+ | 9705 | 4+ | 9709 | 4+ | 9713 | 2+ | 9717 | 4+ | 9721 | 1- |
| 9733 | 1- | 9737 | 4+ | 9741 | 4+ | 9745 | 8+ | 9749 | 3- | 9753 | 2+ | 9757 | 6+ | 9761 | 2+ |
| 9769 | 1- | 9773 | 1- | 9777 | 2+ | 9781 | 1- | 9785 | 4+ | 9789 | 8+ | 9793 | 2+ | 9797 | 2+ |
| 9805 | 3- | 9809 | 4+ | 9813 | 6+ | 9817 | 1- | 9821 | 4+ | 9829 | 3- | 9833 | 3- | 9841 | 8+ |
| 9845 | 4+ | 9853 | 2+ | 9857 | 1- | 9861 | 4+ | 9865 | 2- | 9869 | 2+ | 9877 | 4+ | 9881 | 2- |
| 9885 | 4+ | 9889 | 4+ | 9893 | 2- | 9897 | 6+ | 9901 | 1- | 9905 | 8+ | 9913 | 2+ | 9917 | 2+ |
| 9921 | 2+ | 9929 | 1- | 9933 | 8+ | 9937 | 6+ | 9941 | 1- | 9949 | 1- | 9953 | 4- | 9957 | 2+ |
| 9961 | 2+ | 9965 | 2- | 9969 | 2+ | 9973 | 1- | 9977 | 2+ | 9985 | 2- | 9989 | 2+ | 9993 | 2+ |
| 9997 | 2- | | | | | | | | | | | | | | |

Table 2B

| Disc | H | S | Disc | H | S | Disc | H | S | Disc | H | S | Disc | H | S | Disc | H | S | Disc | H | S | Disc | H | S |
|---|---|---|---|---|---|---|---|---|---|---|---|---|---|---|---|---|---|---|---|---|---|---|---|
| 8 | 1 | - | 12 | 2 | + | 24 | 2 | + | 28 | 2 | + | 40 | 2 | - | 44 | 2 | + | 56 | 2 | + | 60 | 4 | + |
| 76 | 2 | + | 88 | 2 | + | 92 | 2 | + | 104 | 2 | - | 120 | 4 | + | 124 | 2 | + | 136 | 4 | + | 140 | 4 | + |
| 152 | 2 | + | 156 | 4 | + | 168 | 4 | + | 172 | 2 | + | 184 | 2 | + | 188 | 2 | + | 204 | 4 | + | 220 | 4 | + |
| 232 | 2 | + | 236 | 2 | - | 248 | 2 | + | 264 | 4 | + | 268 | 4 | + | 280 | 4 | + | 284 | 2 | + | 296 | 2 | - |
| 312 | 4 | + | 316 | 6 | + | 328 | 4 | - | 332 | 2 | + | 344 | 2 | + | 348 | 4 | + | 364 | 4 | + | 376 | 2 | + |
| 380 | 4 | + | 408 | 4 | + | 412 | 2 | + | 424 | 2 | - | 428 | 2 | + | 440 | 4 | + | 444 | 4 | + | 456 | 4 | + |
| 460 | 4 | + | 472 | 2 | + | 476 | 4 | + | 488 | 2 | - | 492 | 4 | + | 508 | 2 | + | 520 | 4 | - | 524 | 2 | + |
| 536 | 2 | + | 552 | 4 | + | 556 | 2 | + | 568 | 6 | + | 572 | 4 | + | 584 | 4 | + | 604 | 2 | + | 616 | 4 | + |
| 620 | 4 | + | 632 | 2 | + | 636 | 4 | + | 652 | 2 | + | 664 | 2 | + | 668 | 2 | + | 680 | 6 | - | 696 | 4 | + |
| 712 | 4 | + | 716 | 2 | + | 728 | 4 | + | 732 | 4 | + | 744 | 4 | + | 748 | 4 | + | 760 | 4 | + | 764 | 2 | + |
| 776 | 4 | + | 780 | 8 | + | 796 | 2 | + | 808 | 2 | - | 812 | 4 | + | 824 | 2 | + | 840 | 8 | + | 844 | 2 | + |
| 856 | 2 | + | 860 | 4 | + | 872 | 2 | - | 876 | 8 | + | 888 | 4 | + | 892 | 6 | + | 904 | 8 | - | 908 | 2 | + |
| 920 | 4 | + | 924 | 8 | + | 940 | 12 | + | 952 | 4 | + | 956 | 2 | + | 984 | 4 | + | 988 | 4 | + | 1004 | 2 | + |
| 1016 | 6 | + | 1020 | 8 | + | 1032 | 4 | + | 1036 | 4 | + | 1048 | 2 | + | 1052 | 2 | + | 1064 | 4 | + | 1068 | 4 | + |
| 1084 | 2 | + | 1096 | 4 | - | 1112 | 2 | + | 1128 | 4 | + | 1132 | 2 | + | 1144 | 4 | + | 1148 | 4 | + | 1160 | 4 | - |
| 1164 | 8 | + | 1180 | 4 | + | 1192 | 2 | - | 1196 | 4 | + | 1208 | 2 | + | 1212 | 4 | + | 1228 | 2 | + | 1240 | 4 | + |
| 1244 | 2 | + | 1256 | 2 | - | 1272 | 4 | + | 1276 | 4 | + | 1288 | 8 | + | 1292 | 8 | + | 1304 | 6 | + | 1308 | 4 | + |
| 1320 | 8 | + | 1324 | 2 | + | 1336 | 2 | + | 1340 | 4 | + | 1356 | 4 | + | 1384 | 6 | - | 1388 | 2 | + | 1416 | 4 | + |
| 1420 | 4 | + | 1432 | 2 | + | 1436 | 6 | + | 1448 | 2 | - | 1464 | 4 | + | 1468 | 2 | + | 1480 | 4 | - | 1484 | 4 | + |
| 1496 | 4 | + | 1516 | 2 | + | 1528 | 2 | + | 1532 | 2 | + | 1544 | 4 | + | 1560 | 8 | + | 1564 | 4 | + | 1576 | 2 | - |
| 1580 | 4 | + | 1592 | 2 | + | 1596 | 16 | + | 1608 | 4 | + | 1612 | 4 | + | 1624 | 4 | + | 1628 | 4 | + | 1640 | 8 | + |
| 1644 | 4 | + | 1660 | 4 | + | 1672 | 4 | + | 1676 | 2 | + | 1688 | 2 | + | 1704 | 4 | + | 1708 | 12 | + | 1720 | 4 | + |
| 1724 | 2 | + | 1736 | 8 | + | 1740 | 8 | + | 1752 | 8 | + | 1756 | 10 | + | 1768 | 8 | - | 1772 | 6 | + | 1784 | 2 | + |
| 1788 | 4 | + | 1804 | 4 | + | 1816 | 2 | + | 1820 | 8 | + | 1832 | 2 | - | 1848 | 8 | + | 1852 | 2 | + | 1864 | 4 | + |
| 1868 | 2 | + | 1880 | 4 | + | 1884 | 4 | + | 1896 | 4 | + | 1912 | 2 | + | 1916 | 2 | + | 1928 | 4 | + | 1932 | 8 | + |
| 1948 | 2 | + | 1964 | 2 | + | 1976 | 4 | + | 1992 | 4 | + | 1996 | 10 | + | 2008 | 2 | + | 2012 | 2 | + | 2024 | 12 | + |
| 2040 | 8 | + | 2044 | 4 | + | 2056 | 8 | + | 2060 | 4 | + | 2072 | 4 | + | 2076 | 4 | + | 2092 | 2 | + | 2104 | 2 | + |
| 2108 | 4 | + | 2120 | 4 | - | 2136 | 4 | + | 2140 | 4 | + | 2152 | 2 | - | 2168 | 2 | + | 2172 | 4 | + | 2184 | 8 | + |
| 2188 | 2 | + | 2204 | 4 | + | 2216 | 2 | - | 2220 | 8 | + | 2236 | 4 | + | 2248 | 4 | + | 2252 | 2 | + | 2264 | 2 | + |
| 2280 | 8 | + | 2284 | 2 | + | 2296 | 12 | + | 2316 | 8 | + | 2328 | 8 | + | 2332 | 4 | + | 2344 | 2 | - | 2348 | 2 | + |
| 2360 | 4 | + | 2364 | 4 | + | 2380 | 8 | + | 2392 | 4 | + | 2396 | 2 | + | 2408 | 4 | + | 2424 | 2 | + | 2428 | 2 | + |
| 2440 | 4 | - | 2444 | 4 | + | 2456 | 2 | + | 2460 | 8 | + | 2472 | 4 | + | 2476 | 2 | + | 2488 | 2 | + | 2492 | 4 | + |
| 2504 | 4 | - | 2508 | 8 | + | 2524 | 2 | + | 2536 | 2 | - | 2540 | 4 | + | 2552 | 4 | + | 2568 | 4 | + | 2572 | 2 | + |
| 2584 | 16 | + | 2588 | 2 | + | 2604 | 8 | + | 2616 | 4 | + | 2620 | 4 | + | 2632 | 8 | + | 2636 | 6 | + | 2648 | 2 | + |
| 2652 | 8 | + | 2668 | 4 | + | 2680 | 4 | + | 2684 | 4 | + | 2696 | 8 | + | 2712 | 4 | + | 2716 | 4 | + | 2728 | 4 | + |
| 2732 | 2 | + | 2748 | 2 | + | 2760 | 8 | + | 2764 | 2 | + | 2776 | 2 | + | 2780 | 4 | + | 2792 | 2 | - | 2796 | 4 | + |
| 2812 | 4 | + | 2824 | 8 | + | 2828 | 4 | + | 2840 | 4 | + | 2856 | 8 | + | 2860 | 8 | + | 2872 | 2 | + | 2876 | 2 | + |
| 2892 | 8 | + | 2908 | 10 | + | 2920 | 12 | - | 2924 | 8 | + | 2936 | 2 | + | 2956 | 2 | + | 2968 | 4 | + | 2972 | 2 | + |
| 2984 | 2 | - | 3004 | 2 | + | 3016 | 4 | - | 3020 | 4 | + | 3032 | 2 | + | 3036 | 8 | + | 3048 | 4 | + | 3052 | 4 | + |
| 3064 | 2 | + | 3068 | 4 | + | 3080 | 8 | + | 3084 | 4 | + | 3112 | 2 | - | 3116 | 4 | + | 3128 | 4 | + | 3144 | 12 | + |
| 3148 | 4 | + | 3160 | 4 | + | 3164 | 8 | + | 3176 | 2 | - | 3180 | 8 | + | 3192 | 8 | + | 3196 | 16 | + | 3208 | 4 | + |
| 3212 | 4 | + | 3224 | 4 | + | 3228 | 4 | + | 3244 | 2 | + | 3256 | 4 | + | 3260 | 4 | + | 3272 | 4 | - | 3288 | 4 | + |
| 3292 | 2 | + | 3304 | 4 | + | 3308 | 2 | + | 3320 | 4 | + | 3324 | 4 | + | 3336 | 4 | + | 3340 | 4 | + | 3352 | 2 | + |
| 3356 | 6 | + | 3368 | 6 | - | 3372 | 4 | + | 3404 | 4 | + | 3416 | 4 | + | 3432 | 8 | + | 3436 | 2 | + | 3448 | 2 | + |
| 3452 | 2 | + | 3464 | 4 | + | 3480 | 16 | + | 3484 | 4 | + | 3496 | 12 | + | 3512 | 2 | + | 3516 | 4 | + | 3532 | 2 | + |
| 3544 | 2 | + | 3548 | 2 | + | 3560 | 8 | + | 3576 | 12 | + | 3580 | 12 | + | 3592 | 12 | + | 3596 | 12 | + | 3608 | 4 | + |
| 3612 | 8 | + | 3624 | 12 | + | 3628 | 2 | + | 3640 | 16 | + | 3644 | 2 | + | 3656 | 4 | - | 3660 | 8 | + | 3676 | 2 | + |
| 3688 | 2 | - | 3692 | 4 | + | 3704 | 2 | + | 3720 | 8 | + | 3736 | 6 | + | 3740 | 8 | + | 3752 | 4 | + | 3756 | 8 | + |
| 3768 | 4 | + | 3772 | 8 | + | 3784 | 4 | + | 3788 | 2 | + | 3804 | 4 | + | 3820 | 4 | + | 3832 | 2 | + | 3836 | 8 | + |
| 3848 | 4 | - | 3864 | 8 | + | 3868 | 2 | + | 3880 | 4 | - | 3884 | 2 | + | 3896 | 2 | + | 3912 | 4 | + | 3916 | 8 | + |
| 3928 | 10 | + | 3932 | 2 | + | 3944 | 4 | - | 3948 | 8 | + | 3964 | 2 | + | 3976 | 16 | + | 3980 | 4 | + | 3992 | 2 | + |
| 4008 | 4 | + | 4012 | 8 | + | 4024 | 2 | + | 4028 | 4 | + | 4040 | 4 | - | 4044 | 8 | + | 4060 | 8 | + | 4072 | 2 | - |
| 4076 | 2 | + | 4088 | 4 | + | 4092 | 16 | + | 4108 | 4 | + | 4120 | 4 | + | 4124 | 2 | + | 4136 | 4 | + | 4152 | 4 | + |
| 4156 | 2 | + | 4168 | 4 | - | 4172 | 4 | + | 4184 | 2 | + | 4188 | 4 | + | 4204 | 2 | + | 4216 | 4 | + | 4220 | 4 | + |
| 4236 | 4 | + | 4252 | 2 | + | 4264 | 4 | - | 4268 | 8 | + | 4280 | 4 | + | 4296 | 4 | + | 4316 | 4 | + | 4328 | 2 | - |

Table 2B

| Disc | H S | Disc | H S | Disc | H S | Disc | H S | Disc | H S | Disc | H S | Disc | H S | Disc | H S |
|---|---|---|---|---|---|---|---|---|---|---|---|---|---|---|---|
| 4344 | 12+ | 4348 | 14+ | 4360 | 12- | 4364 | 6+ | 4376 | 2+ | 4380 | 8+ | 4396 | 4+ | 4408 | 4+ |
| 4412 | 2+ | 4424 | 8+ | 4440 | 8+ | 4444 | 20+ | 4456 | 2- | 4460 | 4+ | 4472 | 4+ | 4476 | 4+ |
| 4488 | 16+ | 4492 | 2+ | 4504 | 10+ | 4520 | 4- | 4524 | 8+ | 4540 | 4+ | 4552 | 4- | 4556 | 8+ |
| 4568 | 2+ | 4584 | 4+ | 4588 | 4+ | 4604 | 2+ | 4616 | 8+ | 4620 | 16+ | 4632 | 8+ | 4636 | 4+ |
| 4648 | 4+ | 4652 | 2+ | 4664 | 4+ | 4668 | 4+ | 4684 | 6+ | 4696 | 2+ | 4712 | 4+ | 4728 | 4+ |
| 4744 | 8+ | 4748 | 2+ | 4760 | 8+ | 4764 | 12+ | 4776 | 4+ | 4780 | 4+ | 4792 | 2+ | 4796 | 4+ |
| 4808 | 4+ | 4812 | 4+ | 4828 | 4+ | 4844 | 12+ | 4856 | 2+ | 4872 | 8+ | 4876 | 4+ | 4888 | 4+ |
| 4892 | 6+ | 4904 | 10- | 4908 | 8+ | 4920 | 8+ | 4924 | 2+ | 4936 | 4+ | 4940 | 8+ | 4952 | 2+ |
| 4956 | 16+ | 4972 | 8+ | 4984 | 4+ | 4988 | 4+ | 5016 | 8+ | 5020 | 4+ | 5032 | 4- | 5036 | 2+ |
| 5048 | 2+ | 5052 | 4+ | 5064 | 4+ | 5068 | 4+ | 5080 | 4+ | 5084 | 8+ | 5116 | 2+ | 5128 | 4+ |
| 5132 | 2+ | 5144 | 2+ | 5160 | 16+ | 5164 | 2+ | 5176 | 14+ | 5180 | 8+ | 5192 | 4+ | 5196 | 16+ |
| 5208 | 8+ | 5212 | 2+ | 5224 | 2- | 5228 | 2+ | 5240 | 4+ | 5244 | 8+ | 5260 | 4+ | 5272 | 2+ |
| 5276 | 2+ | 5288 | 2- | 5304 | 8+ | 5308 | 10+ | 5320 | 8+ | 5336 | 4+ | 5340 | 8+ | 5352 | 4+ |
| 5356 | 12+ | 5368 | 12+ | 5372 | 4+ | 5384 | 4+ | 5388 | 4+ | 5404 | 16+ | 5416 | 2- | 5420 | 4+ |
| 5432 | 4+ | 5448 | 4+ | 5452 | 4+ | 5464 | 2+ | 5468 | 6+ | 5480 | 4- | 5484 | 8+ | 5496 | 4+ |
| 5512 | 4- | 5516 | 4+ | 5528 | 2+ | 5532 | 4+ | 5548 | 8+ | 5560 | 4+ | 5564 | 4+ | 5576 | 8+ |
| 5592 | 4+ | 5596 | 2+ | 5608 | 2- | 5612 | 4+ | 5624 | 12+ | 5628 | 8+ | 5640 | 8+ | 5644 | 8+ |
| 5656 | 4+ | 5660 | 4+ | 5672 | 2- | 5676 | 8+ | 5692 | 2+ | 5704 | 8+ | 5708 | 2+ | 5720 | 8+ |
| 5736 | 4+ | 5740 | 8+ | 5752 | 2+ | 5756 | 2+ | 5768 | 8+ | 5772 | 16+ | 5784 | 16+ | 5788 | 2+ |
| 5804 | 2+ | 5816 | 2+ | 5820 | 8+ | 5836 | 2+ | 5848 | 8+ | 5852 | 8+ | 5864 | 2- | 5884 | 2+ |
| 5896 | 4+ | 5912 | 6+ | 5916 | 8+ | 5928 | 8+ | 5932 | 2+ | 5944 | 10+ | 5948 | 2+ | 5960 | 4- |
| 5964 | 8+ | 5980 | 24+ | 5992 | 4+ | 5996 | 2+ | 6008 | 2+ | 6024 | 4+ | 6028 | 8+ | 6040 | 4+ |
| 6044 | 2+ | 6056 | 2- | 6060 | 8+ | 6072 | 8+ | 6088 | 12- | 6092 | 6+ | 6104 | 4+ | 6108 | 12+ |
| 6124 | 2+ | 6136 | 28+ | 6140 | 4+ | 6152 | 4+ | 6168 | 4+ | 6172 | 2+ | 6184 | 6- | 6188 | 8+ |
| 6204 | 8+ | 6216 | 8+ | 6220 | 4+ | 6232 | 4+ | 6236 | 2+ | 6248 | 4+ | 6252 | 4+ | 6268 | 6+ |
| 6280 | 4- | 6284 | 2+ | 6296 | 2+ | 6312 | 4+ | 6316 | 2+ | 6328 | 8+ | 6332 | 2+ | 6344 | 4- |
| 6360 | 16+ | 6364 | 4+ | 6376 | 2- | 6380 | 8+ | 6392 | 8+ | 6396 | 24+ | 6412 | 4+ | 6424 | 4+ |
| 6428 | 2+ | 6440 | 8+ | 6456 | 4+ | 6460 | 8+ | 6472 | 4- | 6476 | 2+ | 6488 | 2+ | 6492 | 4+ |
| 6504 | 4+ | 6508 | 6+ | 6520 | 4+ | 6524 | 8+ | 6536 | 4+ | 6540 | 8+ | 6556 | 12+ | 6568 | 2- |
| 6572 | 4+ | 6584 | 6+ | 6604 | 4+ | 6616 | 18+ | 6620 | 4+ | 6632 | 2- | 6636 | 8+ | 6648 | 4+ |
| 6652 | 2+ | 6668 | 2+ | 6680 | 4+ | 6684 | 4+ | 6712 | 2+ | 6716 | 8+ | 6744 | 4+ | 6748 | 4+ |
| 6764 | 4+ | 6780 | 8+ | 6792 | 4+ | 6796 | 2+ | 6808 | 4+ | 6812 | 4+ | 6824 | 2- | 6828 | 4+ |
| 6844 | 4+ | 6856 | 12- | 6872 | 2+ | 6888 | 8+ | 6892 | 2+ | 6904 | 2+ | 6908 | 4+ | 6920 | 4- |
| 6924 | 8+ | 6940 | 12+ | 6952 | 4+ | 6956 | 4+ | 6968 | 4+ | 6972 | 16+ | 6988 | 2+ | 7004 | 8+ |
| 7016 | 2- | 7032 | 12+ | 7036 | 2+ | 7048 | 8+ | 7052 | 8+ | 7064 | 10+ | 7068 | 8+ | 7080 | 8+ |
| 7084 | 24+ | 7096 | 2+ | 7112 | 8+ | 7116 | 4+ | 7132 | 2+ | 7144 | 4+ | 7148 | 6+ | 7160 | 4+ |
| 7176 | 8+ | 7180 | 4+ | 7192 | 4+ | 7196 | 4+ | 7208 | 8+ | 7212 | 8+ | 7224 | 24+ | 7228 | 4+ |
| 7240 | 4- | 7244 | 6+ | 7256 | 2+ | 7276 | 4+ | 7288 | 2+ | 7292 | 2+ | 7304 | 4+ | 7320 | 8+ |
| 7324 | 2+ | 7336 | 4+ | 7340 | 4+ | 7352 | 2+ | 7356 | 4+ | 7368 | 4+ | 7372 | 4+ | 7384 | 4+ |
| 7388 | 6+ | 7404 | 12+ | 7420 | 8+ | 7432 | 4+ | 7464 | 12+ | 7468 | 2+ | 7480 | 8+ | 7484 | 2+ |
| 7496 | 4+ | 7512 | 8+ | 7516 | 2+ | 7528 | 6- | 7532 | 4+ | 7544 | 8+ | 7548 | 8+ | 7564 | 4+ |
| 7576 | 2+ | 7580 | 4+ | 7592 | 4- | 7608 | 4+ | 7612 | 4+ | 7624 | 4- | 7628 | 6+ | 7640 | 4+ |
| 7656 | 16+ | 7660 | 4+ | 7672 | 8+ | 7676 | 4+ | 7692 | 4+ | 7708 | 4+ | 7720 | 4- | 7724 | 2+ |
| 7736 | 14+ | 7752 | 8+ | 7756 | 4+ | 7768 | 2+ | 7772 | 4+ | 7784 | 4+ | 7788 | 16+ | 7804 | 2+ |
| 7816 | 12+ | 7820 | 8+ | 7832 | 8+ | 7836 | 4+ | 7852 | 4+ | 7864 | 2+ | 7868 | 8+ | 7880 | 4- |
| 7896 | 8+ | 7912 | 4+ | 7916 | 2+ | 7928 | 2+ | 7932 | 4+ | 7944 | 4+ | 7948 | 6+ | 7960 | 4+ |
| 7964 | 4+ | 7976 | 2- | 7980 | 16+ | 7996 | 2+ | 8008 | 8+ | 8012 | 2+ | 8024 | 8+ | 8040 | 16+ |
| 8044 | 2+ | 8056 | 4+ | 8060 | 8+ | 8072 | 4+ | 8076 | 16+ | 8088 | 8+ | 8104 | 14- | 8108 | 10+ |
| 8120 | 8+ | 8124 | 20+ | 8140 | 16+ | 8152 | 2+ | 8156 | 2+ | 8168 | 2- | 8184 | 8+ | 8188 | 4+ |
| 8204 | 4+ | 8216 | 4+ | 8220 | 24+ | 8236 | 4+ | 8248 | 2+ | 8252 | 2+ | 8264 | 4+ | 8268 | 8+ |
| 8284 | 4+ | 8296 | 4- | 8312 | 2+ | 8328 | 4+ | 8332 | 2+ | 8344 | 4+ | 8348 | 2+ | 8360 | 8+ |
| 8364 | 8+ | 8376 | 4+ | 8380 | 4+ | 8392 | 4+ | 8396 | 6+ | 8408 | 2+ | 8412 | 4+ | 8440 | 4+ |
| 8444 | 2+ | 8456 | 8+ | 8472 | 12+ | 8476 | 4+ | 8488 | 2- | 8492 | 4+ | 8504 | 2+ | 8508 | 4+ |
| 8520 | 8+ | 8524 | 2+ | 8536 | 16+ | 8540 | 8+ | 8552 | 2- | 8556 | 24+ | 8572 | 6+ | 8584 | 4- |
| 8588 | 4+ | 8616 | 4+ | 8620 | 4+ | 8632 | 4+ | 8636 | 8+ | 8648 | 8+ | 8652 | 8+ | 8668 | 4+ |

Table 2B

| Disc | H S | Disc | H S | Disc | H S | Disc | H S | Disc | H S | Disc | H S | Disc | H S | Disc | H S |
|------|-----|------|-----|------|-----|------|-----|------|-----|------|-----|------|-----|------|-----|
| 8680 | 24+ | 8684 | 4+ | 8696 | 2+ | 8716 | 2+ | 8728 | 2+ | 8732 | 4+ | 8744 | 2- | 8760 | 8+ |
| 8764 | 4+ | 8776 | 4+ | 8780 | 4+ | 8792 | 4+ | 8796 | 4+ | 8808 | 4+ | 8812 | 2+ | 8824 | 2+ |
| 8828 | 6+ | 8840 | 8- | 8844 | 16+ | 8860 | 4+ | 8872 | 2- | 8876 | 4+ | 8888 | 4+ | 8904 | 8+ |
| 8908 | 4+ | 8920 | 12+ | 8924 | 4+ | 8936 | 2- | 8940 | 8+ | 8952 | 4+ | 8956 | 2+ | 8968 | 4+ |
| 8972 | 2+ | 8984 | 2+ | 8988 | 8+ | 9004 | 14+ | 9020 | 8+ | 9032 | 4- | 9048 | 8+ | 9052 | 12+ |
| 9064 | 4+ | 9068 | 2+ | 9080 | 4+ | 9084 | 4+ | 9096 | 4+ | 9112 | 8+ | 9116 | 4+ | 9128 | 4+ |
| 9132 | 4+ | 9148 | 2+ | 9160 | 4- | 9164 | 4+ | 9176 | 4+ | 9192 | 12+ | 9208 | 14+ | 9224 | 8+ |
| 9228 | 8+ | 9240 | 16+ | 9244 | 2+ | 9256 | 4- | 9260 | 4+ | 9272 | 4+ | 9276 | 4+ | 9292 | 4+ |
| 9304 | 2+ | 9308 | 4+ | 9320 | 4- | 9336 | 4+ | 9340 | 20+ | 9352 | 8+ | 9356 | 2+ | 9368 | 2+ |
| 9372 | 8+ | 9384 | 8+ | 9388 | 2+ | 9404 | 2+ | 9416 | 4+ | 9420 | 8+ | 9436 | 8+ | 9448 | 10- |
| 9452 | 4+ | 9480 | 8+ | 9484 | 2+ | 9496 | 2+ | 9512 | 2+ | 9516 | 4- | 9528 | 16+ | 9532 | 2+ |
| 9544 | 4+ | 9548 | 8+ | 9560 | 4+ | 9564 | 4+ | 9580 | 4+ | 9592 | 4+ | 9596 | 10+ | 9608 | 8- |
| 9624 | 4+ | 9628 | 4+ | 9640 | 16+ | 9644 | 2+ | 9656 | 4+ | 9660 | 16+ | 9672 | 8+ | 9676 | 24+ |
| 9688 | 4+ | 9692 | 2+ | 9704 | 2- | 9708 | 4+ | 9724 | 8+ | 9736 | 16+ | 9740 | 4+ | 9752 | 4+ |
| 9768 | 8+ | 9772 | 4+ | 9784 | 2+ | 9788 | 2+ | 9804 | 16+ | 9816 | 8+ | 9820 | 4+ | 9832 | 2- |
| 9836 | 6+ | 9848 | 2+ | 9852 | 4+ | 9868 | 14+ | 9880 | 8+ | 9884 | 4+ | 9896 | 2- | 9912 | 8+ |
| 9916 | 4+ | 9928 | 8- | 9932 | 4+ | 9944 | 8+ | 9948 | 4+ | 9960 | 16+ | 9964 | 4+ | 9976 | 4+ |
| 9980 | 12+ | 9992 | 8+ | | | | | | | | | | | | |

# Index

CPSIA information can be obtained at www.ICGtesting.com

9 780387 970370